Science

When science specialists decide they want to teach, it can be a daunting prospect having to enter the classroom no matter how much subject knowledge they already possess. *Science: Teaching School Subjects 11–19* puts subject knowledge into perspective and shows new teachers and trainee teachers how to make science accessible and interesting for their students.

The book examines the theory and practice of teaching science and includes:

- Science as a school subject
- Teaching science in the secondary school
- Reflecting on science learning – the place of educational research
- Developing science teachers' professional skills
- Future developments in school science.

This book offers a guide for the intellectually alive teacher into the nature of the subject in its entirety and how to think about science when preparing to teach.

Vanessa Kind is Deputy Director, Science Learning Centre North East, University of Durham. **Keith Taber** is Lecturer in Science Education and Programme Manager for the Part-Time Ph.D. in Education at the University of Cambridge.

D0185412

Teaching School Subjects 11–19 Series

Series Editors: John Hardcastle and David Lambert

Mathematics
Candia Morgan, Anne Watson, Clare Tikly

English
John Hardcastle

Geography
John Morgan and David Lambert

Science
Vanessa Kind and Keith Taber

Modern Foreign Languages
Edited by Norbert Pachler, Michael Evans and Shirley Anne Lawes

Business, Economics and Enterprise
Edited by Peter Davies and Jacek Brant

Science:
Teaching School
Subjects 11–19

Vanessa Kind and
Keith S. Taber

Routledge
Taylor & Francis Group

LONDON AND NEW YORK

First published 2005
by Routledge
2 Park Square, Milton Park, Abingdon, Oxon OX14 4RN

Simultaneously published in the USA and Canada
by Routledge
270 Madison Ave, New York, NY 10016

Routledge is an imprint of the Taylor & Francis Group

© 2005 Vanessa Kind and Keith Taber

Typeset in Sabon and Bell Gothic by
Florence Production Ltd, Stoodleigh, Devon
Printed and bound in Great Britain by
MPG Books Ltd, Bodmin

British Library Cataloguing in Publication Data
A catalogue record for this book is available from the British
Library

Library of Congress Cataloging in Publication Data
A catalog record for this book has been requested

ISBN 0–415–36358–6 (hbk)
ISBN 0–415–34283–X (pbk)

Contents

List of illustrations   vii
Series editors' preface   ix
Preface   xv
Acknowledgements   xvii
Glossary   xviii

1   Introduction: an invitation to be a professional science
         teacher   1

Part I   Science as a school subject   **13**
2   What is science?   19
3   What could school science become?   39
4   New perspectives on science education   62
5   What does a science teacher's expert knowledge look like?   92

Part II   Teaching science in the secondary school   **111**
6   Planning to teach the curriculum   114
7   Planning to teach the science   134
8   Acting to teach science   158
9   Science teachers' pedagogic content knowledge   179
10   Evaluating teaching and learning   197

Part III Reflecting on science learning: the place of educational
         research   **209**
11   Being a reflective practitioner   211
12   Using research-based evidence in teaching   224
13   Doing action research in science education   242

v

CONTENTS

Coda: moving science teachers and school science
 forward   257
References   261
Index   271

Illustrations

FIGURE

2.1 Ionic bonding? 30

TABLES

1.1 Key Stages of the English and Welsh system 10
2.1 Science topics in the National Curriculum 22
4.1 Units in the QCA Scheme at KS3 71
4.2 Modules in 'Twenty-first-century science' GCSE courses 75
4.3 Salters Advanced Science units 83
4.4 GCE Applied Science units 87
6.1 Extracts from the Sc1 level descriptions 127
6.2 Two perspectives on teaching 129
6.3 Key points about learning 131
7.1 Structure of an EPIC-designed CBL unit 154

BOXES

1 The Royal Society of Chemistry 14
2 The Association for Science Education 15
2.1 Scientific enquiry in the KS3 National Strategy 34
2.2 Values and science education 37
3.1 Recommendations from 'Beyond 2000' 54
4.1 Purposes of the Framework for teaching science 64
4.2 Guidance on judging progression with the NC levels 67
4.3 KS3 strategy – expectations of schools 69
4.4 Advancing Physics 85
4.5 A constructivist research project in science education 90

ILLUSTRATIONS

5.1	The structure of the secondary curriculum	96
5.2	Teaching across the whole of science	99
5.3	Supporting Physics Teaching 11–14	100
5.4	National Curriculum Statement on Force and Motion	103
6.1	The General Teaching Council for England	118
6.2	Meaningful learning	124
6.3	Bloom's taxonomy	126
7.1	The periodic table	138
7.2	Electronic discussion lists for science teachers	152
8.1	Two teachers acting to teach science	160
9.1	Writing frames	186
11.1	Educational activities of some of the scientific societies	214
11.2	Some approaches to educational research	218
11.3	Drawing conclusions from educational research	220
11.4	Web-based strategies for accessing science education research	221
12.1	Strategies and techniques used in education research	226
13.1	Action research in CLiSP	245
13.2	The Schools–University Partnership for Educational Research	248
13.3	The Collaborative Action Research Network	250
13.4	The British Educational Research Association	251
13.5	Able Pupils Experiencing Challenging Science	252
13.6	KS3 Strategy 'Science Enrichment Project' – on ideas and evidence	255

Series editors' preface

This series aims to make sense of school subjects for new teachers at a moment when subject expertise is being increasingly linked to the redefinition of teachers' responsibilities (Furlong *et al.*, 2000). We start from the common assumption that teachers' passion for their subject provides the foundation for effective teaching, but we also take the view that effective teachers develop a complicated understanding of students' learning. Therefore, we also aim to offer subject specialists a picture of students' learning in their chosen field.

The central argument of the series as a whole is that teachers' professional development in subject specialisms turns on their growing appreciation of the complexities of learning. In essence, the subject knowledge that new teachers bring from their experiences in higher education has to be reworked before it can be taught effectively to children. Our contention is that it is the sustained engagement with the dynamics of students' learning that uniquely sheds light on the way that existing subject knowledge has to be reconfigured locally if it is to be taught successfully in schools. What teachers know about their subject has to be reworked on site, and such is teachers' agency that they will always have a key role in shaping curriculum subjects.

Teaching involves a critical re-engagement with existing subject knowledge. This occurs chiefly through contact with children and communities. All new teachers have to learn how to make complicated judgements about the selection, ordering and presentation of materials with particular learners, real children, in mind. Teachers, then, are learners too. So, as well as giving a picture of students' learning, the series aims to offer a sufficiently complicated account of professional development for new teachers to recognise themselves as learners as they take on new responsibilities in their schools. Thus, we aim to offer insights into the kind of thinking – intellectual work – that teachers at the early stages are going to have to do.

This series is aimed chiefly at new teachers in their years of early professional development. This includes teachers in their initial training year, their induction year and those in years two and three of a teaching career. In addition to Post-Graduate Certificate of Education (PGCE) students and newly qualified teachers (NQTs) working toward the induction standards, the series therefore also addresses subject leaders in schools who have mentor responsibilities with early career teachers, and Advanced Skills Teachers (ASTs) undertaking subject specialist in-service training and teaching support.

The books in the series cover the training standards for NQT status and the induction standards. They use both the training terminology and structure of the official standards in a way that enables readers to connect the arguments contained in the books with their obligation to demonstrate achievement against performance criteria. Yet the books in the series have the ambition to take readers further than mere 'compliance'. They openly challenge teachers to acknowledge their own agency in interpreting 'competence' and to see their role in developing the subject, shaping their professional identities.

A distinctive feature of the series as a whole is its concern with how the particular school subjects have been 'framed'. The books therefore offer a contrast with much that has been published in recent times, including the well-known *Learning to Teach* series, also published by Routledge. They include substantial material on how school subjects connect with wider disciplines, and are also alert to broad social and cultural realities. Thus, they form a response to what has been identified as a major weakness in training and teacher support in recent years – namely, its preoccupation with generic matters of teaching competence at the expense of paying adequate attention to particular issues associated with subject specialism. The books in the *Teaching School Subjects 11–19* series aim to redress the balance.

Those who believe that there is a general 'science' of teaching have been especially influential in recent years. There is no denying that the Key Stage Three Strategy, for instance, has had an impact on the preparedness of teachers generally. Further to this, the identification and recommendation of specific teaching approaches and techniques has enhanced new teachers' technical proficiency generally. Recently, much has been made of teaching 'thinking skills', and such initiatives have raised teachers' all-round performance as well as their professional self-esteem. But when push comes to shove, teaching cannot be sustained in this way. Pupils cannot be taught simply to think. They have to have something to think *about*. If this 'something' is trivial, irrelevant or out of date, then the educational process will be devalued and students will quickly become disaffected. The Secretary of State recognised something of this in 2003 when he launched his *Subject Specialisms Consultation*:

> Our very best teachers are those who have a real passion and enthusiasm for the subject they teach. They are also deeply committed to the learning of their students and use their enthusiasm for their subject to motivate them, to bring their subject alive and make learning an exciting, vivid and enjoyable experience.
>
> It is teachers' passion for their subject that provides the basis for effective teaching and learning. These teachers use their subject expertise to engage students in meaningful learning experiences that embrace content, process and social climate. They create for and with their children opportunities to explore and build important areas of knowledge, and develop powerful tools for learning, within a supportive, collaborative and challenging classroom environment.
>
> (DfES, 2003a: paras 1–2)

The *Teaching School Subjects 11–19* series aims to make practical sense of such assumptions by fleshing them out in terms of teachers' experiences. So, as well as looking at the histories of particular school subjects and current national frameworks, we shall also look at practical matters through case studies and teachers' narratives. We have noted how new teachers sometimes feel at a loss regarding the very subject knowledge they carry forward from their previous educational experiences into teaching. This feeling may be due to their entering a highly regulated profession where it appears that choices concerning what to teach (let alone how to teach it) are heavily constrained. However, much will be lost that could sustain creative and healthy classrooms if the system cuts off a primary source of energy, which is teachers' enthusiasm for their subject. Good teachers connect such enthusiasm with the students' interests. The *Teaching School Subjects 11–19* series engages with just this issue. If it has a single, clear mission, it is to encourage the thought in teachers that they do not merely 'deliver' the curriculum in the form of prefigured subject knowledge, but that they have a genitival role in making it.

What does it mean to 'make' a curriculum? This is a huge question and we do not aim to provide a definitive curriculum theory. However, we note that current accounts of curriculum and pedagogy (e.g. Moore, 2000) tend to emphasise the role of competing interests that decide the educational experience of students. They offer a complicated picture of curriculum construction by taking in societal, economic and cultural influences. Plainly, no single interest wholly determines the outcome. Additionally, there is a growing agreement among educationists in England and Wales that 'central government control of the school curriculum must be loosened' to release teachers' energies (White, 2004: 189). We adopt a position similar to John White's, which is to 'rescue' the curriculum from central prescription and 'to see teachers having a greater role than now in . . . decisions on the curriculum . . . (ibid.: 189–90).

This is not to say that the government has no role at all. Few educationalists would want to return fully to the arrangements before the 1988 Education Reform Act, when the curriculum experience of students was almost entirely in the hands of teachers and other interest groups. It is surely right that the elected government should regulate what is taught, but not that it should prescribe the curriculum in such an inflexible manner that it stifles teachers' initiative. Teachers play an active role in shaping the curriculum. They make professional decisions given, as White puts it, their 'knowledge of the pupils on whom the curriculum will be inflicted'. We argue that it is here, in deciding what to teach and how to teach it, that teachers' knowledge and creativity is of cardinal value. Teaching is quintessentially a practical activity and teachers' performance matters. But we also know that behind the creativity in teaching lies a form of intellectual work. Our starting position is that intellectual effort is required at every stage of teaching and learning if it is to be worthwhile.

Knowledge of the pupils is a fundamental component of curriculum design. Effective teachers are in secure possession of just this kind of knowledge of their pupils where it informs their decisions about the selection of content and the choice of methods. However, the series also makes it plain that knowledge of the pupils on its own is an insufficient basis for working out what to teach and how to teach it. Secure subject knowledge is equally important. Furthermore, we take the view that an essential element of a secondary teacher's professional identity is tied up with a sense of their subject specialism. It is generally true that effective teaching requires a deeper grasp of a subject than that specified in the syllabus. What is more, pupils frequently admire teachers who 'know their stuff'. What 'stuff' means is usually larger than a particular topic or a set of facts. Indeed, the way that an effective teacher makes a particular topic accessible to the pupils and enables them to progress often relies on their having a good grasp of the architecture of the subject, what the main structures are and where the weaknesses lie. You can't mug this up the night before the lesson.

It is widely recognised that PGCE students and early career teachers frequently turn to school textbooks to fill the gaps. This is fine – inevitably there will be aspects of the subject that the specialist has not covered. Many teachers now use the internet proficiently as a rich source of information, data, images and so on, which is also fine. But what teachers also need to do is to make sense of the material, organise it and sift it for accuracy, coherence and meaning. The series helps new teachers to do this by taking them into the relevant subject debates. The authors introduce teachers to the conceptual struggles in the subject and how these impinge on the making of the school subject. Through debating the role of the school subject, and showing how it hangs together (its 'big concepts'), they also show how it contributes to wider educational aims. Such a discussion takes place in the context of renewed debate about the future of school subjects and the subject-based curriculum. Although the series serves the needs of subject specialists, it does not take as given the unchanging status of school subjects, and the authors will take up this debate explicitly.

Current notions of subjects as inert 'contents' to be 'delivered' grate against learning theories, which foreground the role of human agency (teachers and pupils) in the construction of knowledge. For the teacher, good subject knowledge is not about being 'ahead of the students', but being aware of the wider subject. Teachers might ask themselves what kinds of knowledge their subject deals with. And, following on from this, they might also ask about the kinds of difficulties that students often encounter. Note that we are not concerned with 'correcting' pupils' 'misconceptions' about what they get from their lessons, but with what they actually make of what they get.

The series has a broad theoretical position which guides the way that the components of the individual books are configured. These components include lesson planning, classroom organisation, learning management, the assessment of/for learning and ethical issues. However, there is no overarching prescription and the various volumes in the series take significantly different approaches. Such differences will depend on the various priorities and concerns associated with particular specialist subjects. In essence, the books aim to develop ways of thinking about subjects, even before teachers set foot inside the classroom.

We doubt the adequacy of any model of teaching and learning that reduces the role of the teacher to that of the technician. Teachers mediate the curriculum for their students. Furthermore, there is an urgent justification for this series of books.

It is the ambition of the series to restate the role of subjects in schools, but not in a conservative spirit that fails to engage with substantial change and developments. For some commentators, the information explosion, together with the still-quickening communications revolution, spells the death of subjects, textbooks and the rest of the nineteenth-century school apparatus. Although we do not share this analysis, we acknowledge that the status quo is not an option. Indeed, subject teachers may need to become less territorial about curriculum space, more open to collaboration across traditional subject boundaries and more responsive to what have been called 'unauthorised subject stories' – student understandings, media representations and common-sense views of the world. In such an educational environment, we would argue, the role of disciplinary knowledge is even more important than it was a decade ago, and teachers need to engage with it creatively.

The *Teaching School Subjects 11–19* series aims to support new teachers by helping them to discover productive ways of thinking about their specialism. The specialist authors have tried to maintain an optimistic, lively and accessible tone and we hope you enjoy them.

John Hardcastle and David Lambert
London, 2004

Preface

This book about science as a school subject is for people thinking about, planning to become, or who are in the process of becoming, science teachers, as well as their advisers, mentors and tutors. We think of our audience as experts in some aspect of natural or applied science. Therefore, we may seem to be writing a relatively short book about science for people who already know about science. In fact, this book is not about the science that our audience already knows. The rationale for our book can be explained in two related premises that underpin everything that follows. These are that:

- science teachers are not scientists;
- school science is not science.

Clearly, we will qualify both of these statements. However, we believe that engaging with these statements is central to becoming an effective science teacher.

We will argue that the process of becoming an effective science teacher requires taking on a new professional identity and adapting to new specialist knowledge. Knowing and being able to 'do' some science are prerequisites for becoming a science teacher, but these are not the only requirements of science teaching. Similarly, school science obviously has some relationship with the body of knowledge considered to be 'science'; yet this relationship is complex and imperfect.

This book aims to provoke the reader to engage critically with the key issues of what it means to be a science teacher and the science that teachers are expected to teach. In exploring these issues, key concepts and principles will be introduced, and the work of some seminal thinkers and researchers will be name-checked. The reader will also be introduced to aspects of the current 'system' within which science teachers work. We hope this book will provide a useful introduction to these important areas, but we have not written a detailed textbook about them. We see our job as being to help the reader to see how and why these things matter in the overall process of becoming a professional science teacher.

We also have a vision of the kind of science teacher we would like to develop. Both of us have observed many trainees on school placements and worked with a wide variety of practitioners of varying degrees of experience and competence. We claim that to be truly effective as a science teacher requires the provision of challenging, appropriate and stimulating experiences for our learners and, by default, an understanding of what such experiences may be for the learners being taught. We are fully aware that making this happen presents a challenge. We have written our book firmly in the belief that school and college students deserve the absolute best a science teacher can provide for them and hope that readers will be inspired and challenged by what we say.

Vanessa Kind and Keith Taber, 2005

Acknowledgements

We thank David Lambert, Series Editor, for his patience and support during the writing process. We also record our thanks to the trainees we have met, worked with, observed and helped to induct in the process of becoming professional science teachers. Our work with them gave us the inspiration for many of the beliefs we record here. We also need to thank our families who have tolerated the periods of obsessive preoccupation that are involved in the production of any book.

Glossary

The UK education scene is full of abbreviations and acronyms. Therefore, we think it helpful to define and explain the most commonly used ones at the outset, as well as setting out terms that we will use in particular ways. Sometimes this usage is primarily for our convenience and to maintain a straightforward writing style, but we also wish to emphasise certain terms reflecting some of our key messages.

Specific terms we use are:

inductee A person in the process of changing professional identity from a specialist 'scientist' or other role to 'science teacher', such as a trainee teacher or NQT.

maintained schools Schools funded from public money, expected to follow government education policies.

school Secondary schools (11–16, 11–18 and 13/14–19 institutions) sixth-form colleges (post-compulsory college for 16–19-year-olds), further education (FE) colleges (post-compulsory college offering a wide range of post-16, diploma and sometimes degree courses).

science We will use the term SCIENCE (in small capital letters) to represent the activity undertaken by professional scientists to distinguish it from the school subject called 'science'. School science is intended to reflect SCIENCE, but we think it necessary to keep in mind the limitations of the comparison. (Where we discuss the general body of knowledge labelled science, or a teacher or students' knowledge of science, we will use lower case type.)

scientist Someone working as a professional scientist or engineer. Although many teachers of science consider themselves scientists we will refer to them as science teachers.

student   A learner attending a school or college. Note that 'student teachers' are now usually referred to as 'trainees'.

trainee   Someone following a course of professional training as a teacher. 'Student teacher' is no longer commonly used as trainees are considered to be professionals-in-training and may receive a 'training salary'. Attendance at all sessions at a Higher Education Institution (HEI) (see p. xx) and school as part of the training are considered mandatory for trainees in the same way as for a contracted employee.

The following are common acronyms or expressions used in science teaching and education in the UK:

A level   Advanced level: the most common qualification taken for university entrance, usually taken over two years. The first year comprises AS courses and the second year A2 courses.

AS level   Advanced Supplementary level: a one-year, post-16 qualification making up the first half of an A level qualification sometimes referred to as 'A1'.

A2 level   Advanced 2 level: a one-year, post-16 qualification making up the second half of an A level qualification.

ASE   The Association for Science Education: a professional organisation for science teachers.

AT   Attainment Target: part of the National Curriculum specifying the level of knowledge students are expected to reach at each Key Stage.

BA or BAAS   British Association for the Advancement of Science – an organisation open to scientists and non-scientists with an interest in science. The BA organises an annual science festival which is well attended (including by teachers) and well covered in the new media.

CPD   Continuing Professional Development: activities that ensure that a teacher (or other professional) continues to develop and update their professional knowledge and skills after initial qualification.

CSE   Certificate of Secondary Education: an examination that used to be taken by 16-year-olds as an alternative to O level. O level and CSE examinations were merged to form the GCSE (see p. xx) from 1989.

FE   Further Education: a term used to describe non-university post-compulsory education taken after age 18. FE colleges specialise in this area.

GCE   General Certificate of Education: the formal term used to describe A levels. (Previously this included O level as well as A level.)

GCSE General Certificate of Secondary Education: the school examination usually taken at age 16 that marks the end of compulsory education in England, Wales and Northern Ireland.

GNVQ General National Vocational Qualification: a two-year, usually full-time, job-related course offered in sixth form and FE colleges for students wishing to undertake training leading to employment in specific fields such as Leisure and Tourism, Health and Social Care and Science. GNVQs have now become GCSEs in vocational subjects and VCEs (see p. xxii).

GTCE The General Teaching Council for England: the professional regulatory body for teachers working in England.

GTP Graduate Teacher Programme: an employment-based route into teaching used mainly by mature entrants and those returning to work after a career break.

GTTR The Graduate Teacher Training Registry: the clearing house handling applications to PGCE courses in the UK.

HEI Higher Education Institution: in our context an establishment, usually a university, offering initial teacher training courses.

InSET In-service Education for Teachers: further 'training' once qualified and in post. Although many teachers still use the term InSET, providers tend to use the term CPD.

IoB The Institute of Biology: the professional body and learned society for biological scientists.

IoP The Institute of Physics: the professional body and learned society for physical scientists.

ITT Initial Teacher Training: the course of training leading, over two years, to Qualified Teacher Status.

KS3 Key Stage 3: ages 11–14 (years 7–9) in the English and Welsh system. Note that KS1 and KS2 also exist, but these apply to primary aged children and therefore are outside the scope of this text (see Notes on the national context, pp. 8–11).

KS4 Key Stage 4: ages 14–16 (years 10–11) in the English and Welsh system.

KS5 Key Stage 5: a term sometimes used for post-16 (years 12–13) studies, usually A level, in the English and Welsh system.

LEA Local Education Authority: an administrative organisation responsible for education within a specified area of the UK.

NC National Curriculum: the statutory curriculum taught in maintained schools.

NQT Newly qualified teacher: someone who has met QTS standards and is undergoing a structured induction period, usually over one year, in their first teaching post.

Ofsted The Office for Standards in Education: an official body that inspects and reports on schools and other institutions involved in education.

O level Ordinary level: examinations taken at 16 prior to and up to 1988. O levels were merged with CSEs to create GCSEs.

PGCE Post-graduate Certificate in Education: the standard teaching qualification in the UK gained by following a one-year, full-time course at an HEI. A prerequisite is possession of a Bachelor degree, hence 'post-graduate'. (In some HEIs, PGCE stands for *Professional* Graduate Certificate rather than Post-graduate. The Post-graduate label implies study substantially at Masters degree level.)

PoS Programme of Study: the NC document specifying the content to be taught in a subject.

QCA The Qualifications and Curriculum Authority: the organisation overseeing the standard of examinations and SATs in England and Wales.

QTS Qualified Teacher Status: the formal status awarded by the UK Government on the recommendation of HEIs to those considered to have met the 'teaching standards' during a period of teacher training, such as a PGCE.

RSC Royal Society of Chemistry: the professional body and learned society for chemical scientists.

SAT Standard Assessment Test: a test set by government to measure children's achievements at the ends of Key Stages 1, 2 and 3 in English, mathematics and science. At KS4 GCSEs provide the final assessment.

SCITT School Centred Initial Teacher Training: a course leading to Qualified Teacher Status directed by a group of schools working together with relatively little involvement from an HEI.

Selective school A secondary school that decides on its own student intake, either by examination and/or interview, or from evidence of other talent such as drama or music. Selective schools identifying children by IQ examination are now comparatively rare in England.

Sixth form Term used to describe years 12 and 13, for 16–18-year-olds, in a secondary school. This developed from the old way of numbering year groups from the start of secondary school – '1st form' being 11-year-olds and so on. This was replaced in the late 1980s by a system numbering years from the beginning of primary schooling at age 5, but the term 'sixth form' is still used in many schools and 'sixth form colleges'.

SoW   Scheme of Work: a teaching plan covering a sequence of lessons for one or more topics.

TTA   The Teacher Training Agency: the organisation overseeing recruitment into teaching in the UK.

VCE   Vocational Certificate of Education: a qualification equivalent to A level, that is, taken after two years post-compulsory study, in a subject and style related to employment. From September 2005 these will have an AS/A2 structure like A levels and have titles such as 'Applied Science'.

Chapter 1

Introduction

An invitation to be a professional science teacher

Children seldom misquote you. In fact, they usually repeat word for word what you shouldn't have said.

(Anonymous)

AN INVITATION

This book is one of a series on teaching school subjects, focusing on teaching science to 11–19-year-olds. The reference to science as a 'school subject' means *more* than teaching science in a school context. SCIENCE (see Glossary for explanation) is something undertaken by professional scientists working in academic and industrial situations. *School science*, by contrast, is a curriculum subject taught by members of the teaching profession in educational institutions. A key theme of this book is the notion that 'school science' is an entity with its own identity, having limited overlap with SCIENCE itself. This means that teaching science in school involves (among other things) developing a familiarity with the entity known as 'school science'. More significantly, for those persuaded by our argument that the relationship between SCIENCE and school science is not straightforward, then 'school science' becomes something:

- requiring justification;
- developing independently of science; and
- open to being moulded by policy-makers.

These characteristics lead us immediately to questions about school science:

1 Why is science worth teaching in schools?
2 What causes school science to develop?

3 Who determines (and who *should* determine) what school
   science is?

Without wishing to spoil any intrigue these questions may create, we
reveal at this early stage that our view of who should be moulding 'school
science' is very much tied to our vision of what being a *professional*
science teacher means. Consider an analogy between teaching and the
practice of SCIENCE.

Imagine working on a practical project where you are given full
training on the use of the laboratory apparatus, and closely supervised
as you learn new techniques. In this scenario, you are given responsibility for selecting specific techniques for answering certain key questions,
but it is not considered your role to know which theoretical considerations drive the research programme, or even why the research is being
undertaken. We feel that such a scenario is actually quite common, except
we would normally consider the person undertaking this type of work
to be a *technician* rather than a *scientist*.

This is in no way meant to denigrate the work of technicians, which
is essential in both science and teaching, but we do not feel science
teachers should be limited to being educational technicians. Echoing the
sentiments of the series editors in their introductory comments, 'we doubt
the veracity of any model of teaching and learning that designates the
teacher in a relatively powerless technician role'.

We do not feel this analogy is fanciful. We all know that some technicians are well trained, highly skilled, very creative within their sphere
of responsibility, are given much due respect by peers and others, and
can be trusted to work with minimal supervision on important work
with precious materials. Yet, if they have no input in determining the
aims of the work, and are not privy to discussions about its meaning
and consequences, they are still undertaking technical rather than professional roles. We feel that some key aspects of the way the education
system is currently structured and judged tend to de-professionalise
teachers and, for that matter, teacher–educators.

We would suggest that the matters discussed in this book are of
fundamental importance to science teaching, and to the development of
science teachers, individually and collectively, as professionals. It is especially important to raise this issue at the start of the teaching career, as
the 'inductee' will naturally expect to be taking advice and instruction
from tutors, mentors, heads of department and so forth. It is important

to realise that once you have earned the title of qualified teacher you should expect to be given a level of responsibility commensurate with that professional status.

This process of becoming a professional science teacher is at the heart of this book. It would be satisfying to suggest that reading this book will enable such a process – but that would be naive, because we do not think that the profession has yet fully grasped what being 'professional' means, and arrogant because few books change lives significantly of themselves. Reading requires critical reflection to lead to significant learning and the resultant thinking has to be integrated into action if the learning is to have a practical outcome. We have a view of what being a professional science teacher means. We will share that with the reader to show how our notion of professionalism is tied to questions such as what school science is, could and should be. We *might* also convince you to share our view of the professional science teacher, but we will settle for critical engagement with the issues as a minimum.

So, we address this book to those who seek to become professional science teachers. We envisage the readership to be largely new entrants to the profession, such as trainees seeking Qualified Teacher Status (QTS) or those working through the newly qualified teacher (NQT) year. However, we hope this book will be read by those exploring what being a professional science teacher means: whether considering teaching as a career, about to embark on teacher training, seeking a new leadership role in a science department, or simply wishing to engage with issues as part of continuing professional development.

EXCLUSIONS – THE SMALL PRINT

Here, we clarify what our book does not do. Reading this book will not tell a reader *all* s/he may need to know about the curriculum, modes of assessment or teaching approaches. We have things that we feel are important to say about these issues, and we will alert you to some key themes and debates, but we will not provide comprehensive accounts of what to teach, how to teach or how to assess. Nor will this book provide an in-depth account of government policies, guidance and initiatives, or details about what to include in a lesson on metals to 14-year-olds or how to set up apparatus to demonstrate convection. (There are some very good books available that do provide detailed discussion on such specifics, and the latest government requirements and advice are readily available from various official websites.) This is because we focus on the central ideas we believe are fundamental to becoming a professional science teacher.

We believe strongly that you should be very clear about exactly why you wish to teach, or are teaching, science and that you have in place solid foundations for your subsequent professional learning. These foundations could be described as an explicit personal philosophy of teaching that will guide your thinking and action as a teacher. This is one aspect of what we mean when we talk about being a professional science teacher.

This book introduces what we consider are fundamental issues for the professional science teacher. We aim to provide a *framework for thinking* about the key areas involved in acting as a science teacher. New entrants are encouraged to *read this book first*, using the ideas presented to inform subsequent (critical and reflective) reading. Our intention is to provide insights that will underpin reading about teaching science throughout a developing teaching career. We hope that, despite changes to curricula, assessment procedures and other educational policies, the foundations we offer will continue to be relevant.

We also expect that most readers will see themselves as scientists or engineers (something we discuss in Part I) used to studying science at a high level. Accordingly, we do not anticipate familiarity with ideas and approaches from the social sciences. We warn you that we are no longer, in a professional sense at least, natural scientists. Our research is published in *education* journals, not SCIENCE journals. Although we hold science degrees and both have considerable experience in teaching science and in helping others to teach science, our professional work has made us into *social* scientists. We think this is a very important transition to appreciate. After all (and this may come as a shock) any inductee is undergoing the same transition.

INCLUSIONS – THE STRUCTURE OF THE BOOK

The book is divided into three parts, intended to reflect stages in the professional development of a science teacher. Loosely, these sections concern the nature of school science, teaching science and researching your own teaching.

The rationale for the first section is simple, following from what we have said above. To be an effective teacher of school science means having a good understanding of what school science is. This does not mean just learning which topics are included in the curriculum, but also understanding the process by which the curriculum comes about, while appreciating the teacher's professional role in forming and interpreting the curriculum.

The second part focuses on teaching science in a secondary school or post-16 college. Although some readers may think this is the most important part of the book, we recommend that this section is best read in the light of the first section. Here, we look at some important aspects of classroom work – such as planning teaching, the place of practical work, assessing students' learning and understanding the learning process. Teachers have to make many important decisions on a daily basis, based on their knowledge and understanding of these areas. To know if these are the right decisions, teachers need to have an underpinning personal philosophy of teaching. This section will discuss three stages in the teaching process, each necessary for overall teaching effectiveness:

- *planning* to teach science;
- *acting* to teach science;
- *evaluating* teaching.

The final part develops the importance of evaluating our professional work, focusing on the need for teachers to reflect upon their practice. This section discusses aspects of educational research that influence the professional science teacher. A key part of our thinking here is that educational research concerns *all* professional teachers in two ways. The first is obvious: professionals have a responsibility to 'keep up to date' by following the latest developments in their field. As a science teacher you will want and need to 'keep up' with major developments in the SCIENCES, so that your reading (and viewing of documentaries, etc.) will inform your teaching. We certainly encourage this, but with the reminder that SCIENCE is no longer your primary professional field. As a professional educator, this brings a responsibility to be aware of important developments relating to the understanding of teaching and learning, as well as to developments in understanding superconductors or gene expression.

This type of keeping up may seem quite difficult for teachers, as although there are many education journals, most are very expensive and not readily accessed in schools (although useful material is becoming increasingly available through the world wide web). However, there are ways of keeping informed through courses, links with Education Faculties in the local universities, and research digests in the educational and quality daily/weekend press.

The second way that we think that educational research should inform the work of professional science teachers is through a personal involvement in the research process. Time and resource constraints do not allow most teachers to undertake substantial research activity, but we believe that all professional teachers should undertake small-scale, classroom-based enquiries within their own contexts. A key here is the notion of 'evidence-based practice': that our actions as professionals should be justified in terms of available evidence. Some of that evidence may come from published research, but other evidence derives from the work of the individual teacher taking an enquiry-based approach in her own classroom.

WHAT ARE WE AIMING FOR?

This book is about what becoming a professional science teacher means in a school or college context. Consider yourself in this context by thinking of the values you wish to bring to your teaching and how you would wish to be remembered by the children or students you teach. Here is an account of an event that happened to one of the authors:

> I taught for a time in the late 1980s at a boys' school in east London. The science course for 11–14-year-olds involved a lot of practical work, including technologically based activities. One, with year 7 (11-year-olds) was building a 'balloon machine'. This was a device to be constructed from assorted lab equipment, string, cardboard and anything else available with the premise that a candle had to be lit to burn a string to set off a chain of events leading to a pin popping a balloon. The boys got deeply involved in the task and came up with some very good, complex machines. We demonstrated the most ingenious at a parents' evening. I really did not give the lesson much thought until years later, when giving a lecture at an A level chemistry conference. The organisers came up to me and said that there were some 'young men' who wanted to meet me. 'Great! A fan club!' I thought, somewhat ironically. I vaguely recognised the enthusiastic group who greeted me with 'Do you remember the *balloon machines*?!' Embarrassed that I could not remember a single one of their names, they went on to re-introduce themselves and described how these early lessons had influenced them and that they had ended up continuing

to study science as a result. It was humbling to realise that they had remembered what to me had been a 'one-off' event and that they perceived me as having had so much influence on them.

One very important point comes out of this story – children remember their teachers, much more than teachers remember the children, hence the quote at the start. The author in question was embarrassed by hardly being able to recall the lesson, let alone a single name of the children, then young men – yet the boys could recall many more details. The impact this simple activity had had on the audience went far beyond what the teacher imagined at the time. As a professional science teacher, therefore, you are in a much more influential job than you perhaps realise. The potentially long-lasting nature of the influence you wield means that the job deserves to be done as well and thoroughly as possible. We aim in this book to set out our personal beliefs about taking on this professional identity, helping you to identify your own philosophy of science teaching, understanding key aspects of the work of science teachers and developing into a reflective practitioner who seeks and evaluates evidence to help adapt your practice. None of this can be achieved simply by reading a book – but we trust this book will guide you through this difficult, exciting and potentially worthwhile process.

THINKING ABOUT PRACTICE

At the end of each chapter we offer questions to prompt you to think more about the issues raised. They may also point to what is to come next. We suggest you discuss your answers and ideas with others – perhaps fellow trainees, or those working outside education. The point is for you to be able to practise formulating and articulating your own thoughts and ideas about science teaching.

- Why is science worth teaching in schools?
- What causes school science to develop?
- Who determines what school science is?
- What would you most appreciate your students remembering about your science teaching?

A NOTE ON THE NATIONAL CONTEXT OF THE BOOK

Our book has (like the others in the series) been written within the context most familiar to the authors, that is the UK and, in particular, the English school system. We provide these notes to help orientate readers from outside the UK to the ways in which we express information about children's ages, school type and so on.

GENERAL DIFFERENCES BETWEEN PARTS OF THE UK

The UK, comprising England, Wales, Scotland and Northern Ireland, does not have a unified education system. In the UK we are used to and appreciate that this complexity accommodates local trends, traditions and conditions, but it is hard to understand at first. Without going into excessive detail, as it would probably be possible to write an entire book to inform about the complete range of differences, we give some examples. Scotland and Northern Ireland differ from each other and from England and Wales in the organisation of their secondary schools in terms of type of school and age of transition between phases. Scotland does not use a statutory National Curriculum (NC) and most Scottish students complete school at age 17. England and Wales have the same statutory NC. The Welsh language has special status in Welsh schools, but is completely absent from all other parts of the UK system. We refer to the UK where our comments are applicable across all the countries of the United Kingdom, but to England or England and Wales where details differ.

THE UK SCHOOL YEAR AND STARTING SCHOOL

International readers should also note for contextual purposes that the school year in all parts of the UK runs generally from the beginning of September to mid- or late July. The long summer holidays therefore start in July and continue through August. The age at which children start school is calculated using the school year, for example, from 1 September 2004–31 August 2005. Children begin compulsory education at age 5, that is, in the school year in which their 6th birthday falls between 1 September and 31 August. However, in a majority of cases, children start formal schooling in the six months or year before this in a 'Reception' or 'Rising 5s' class.

PHASES OF EDUCATION

In England, Wales and Northern Ireland the term 'primary' is used to mean education between ages 5 and 11, 'secondary' for education between 11 and 18 and 'tertiary' for education for post-18-year-olds. Secondary education is compulsory to age 16. Other terms are also used to describe tertiary education: 'further' education (FE) normally refers to post-16 education taking place in colleges that are not universities. 'Higher' education (HE) normally refers to post-18 education taking place in universities and (confusingly) some colleges leading towards a degree qualification.

Although the age divisions between phases are constant, LEAs organise for themselves the age at which children transfer between primary and secondary schools. In most English and Welsh LEAs children transfer from primary to secondary schools at age 11. In some, children attend a 'middle school' from the age of 8/9 and transfer at 12, 13 or 14.

THE NATIONAL CURRICULUM FOR ENGLAND AND WALES

The National Curriculum (NC) states the content that must be taught to children in compulsory education, that is, aged 5–16. The NC document divides compulsory education into four 'Key Stages', as shown in Table 1.1. At the end of each Key Stage, children take tests. In KS1, KS2 and KS3 these are Student Assessment Tests (SATs) and at KS4 GCSEs. The relationship between Key Stage, year and students' ages are shown in Table 1.1.

POST-COMPULSORY EDUCATION

The term 'KS5' is sometimes used to signify the next stage of secondary education. Examination boards specify the content for post-16 courses, working in conjunction with the QCA and DfES. As of yet, there is no statutory NC for post-16 education in England and Wales in terms of which subjects students should study. Hence, students can study a wide range of different subjects, usually taking four or five in the AS (Y12) year reducing to three in the A2 year. Students do not have to study English, maths, science or a language beyond the age of 16. Sixth forms and colleges tend to recommend certain subject combinations and some subjects are regarded as essential if a student wishes to study a certain subject at university. All syllabuses, now officially called 'specifications',

▨ *Table 1.1* *Key Stages of the English and Welsh system*

Key Stage (KS)	Year	Age of students
1	1	5–6
	2	6–7
	3	7–8
2	4	8–9
	5	9–10
	6	10–11
3	7	11–12
	8	12–13
	9	13–14
4	10	14–15
	11	15–16
'KS5' or 'sixth form'	12, also called 'lower sixth' or 'A1'	16–17
	13, also called 'upper sixth' or 'A2'	17–18

have to be scrutinised and agreed by the QCA as meeting the same standards and content in each subject. This ensures that universities can, as far as possible, be assured that all A levels are of the same standard.

Students may need an extra year at school, for example, to re-take examinations because of illness or for some other justifiable reason. Hence, some students complete secondary education at the age of 19 and this is often, but not always, reflected in the descriptions of the age range taught at different schools and colleges.

Compulsory education ends at age 16, so years 12 and 13 are optional. LEAs differ in their organisation of 16–19 education. Some use mainly 11–18/19 secondary schools, so students can 'stay on at school' if they wish to. Most post-16 students attending 11–18 schools take A levels. Other LEAs provide 11–16 or 12–16 secondary schools only. This means that at the end of compulsory education anyone wishing to 'stay on' will need to attend a 'sixth form college' or other further education (FE) college. A sixth form college normally educates students aged 16–19. An FE college will have 16 as a minimum enrolment age and, besides A levels, will also offer a wide range of courses to a local community, including vocational, academic, basic skills and degree courses, as well as non-qualification courses for adult literacy, numeracy, computing and foreign language skills, among others. The normal minimum age for entry to higher education is 18.

Part I

Science as a school subject

In this part, we introduce the subject matter of school science, what school science could become, new perspectives on school science, and give a view about what a science teacher's knowledge about SCIENCE may be. However, we would like to start by inviting readers to consider the process of becoming a science teacher.

New entrants to teaching have varied backgrounds. For example, some enter training immediately after graduating with their bachelor's degrees, or after having a 'gap' year travelling (and sometimes teaching) abroad. Others have completed masters or doctorate degrees. Some new entrants have worked in scientific research, while others have been employed in scientific industries in research, development, sales or management positions. There are other trainees who may not have used their science knowledge for some years, perhaps having taken jobs in which their degree subject was not relevant or who are returning to work after raising a family. Therefore, most new entrants to science teaching come with a professional identity relevant to their previous life. This could be 'biologist', 'electrical engineer', 'materials scientist', 'fireman', 'doctor', 'medical sales representative', 'full-time homemaker', 'management consultant' or 'Disney World entertainment officer'. The more scientific or professional labels may relate to an organisation regulating and defining the professional group. For example, in the UK, the Royal Society of Chemistry is the professional body and learned society for chemical scientists (see Box 1).

As part of the identity 'shift', many join the subject association for teaching science, the Association for Science Education (ASE, see Box 2).

We suggest that part of the process of becoming a science teacher involves internalising the change of professional identity from 'specialist scientist' (or other role) to *science teacher*, making this the primary professional identity. To take this a step further, we (the authors) have multiple professional identities. We both studied science subjects for first degrees. One of us worked in research,

BOX 1 THE ROYAL SOCIETY OF CHEMISTRY

As the professional body for chemistry in the UK, the RSC is responsible for maintaining advanced standards of qualifications, competence and professional practice among chemical scientists. The RSC assesses and accredits degrees and diplomas in the chemical sciences and related courses in British universities.

As a professional qualifying body, the RSC stipulates criteria for membership through four professional categories, and awards Chartered Chemist status.

The RSC plays a leading role in the chemical sciences, communicating cutting-edge research and its applications through highly respected journals and its programme of international conferences, seminars and workshops.

The RSC is the largest organisation in Europe for advancing the chemical sciences. Supported by a network of 45,000 members worldwide and an internationally acclaimed publishing business, our activities span education and training, conferences and science policy, and the promotion of the chemical sciences to the public.

(Text edited from the RSC website www.rsc.org, accessed 6 April 2004)

but we both became teachers and, later, academics. Although we do not think we have ceased to be scientists, when we became *science teachers* our primary professional identities changed. When we moved to working in universities our professional identities changed again, because just as 'teaching science' differs from 'doing SCIENCE', educating science teachers is also different! The one of us who has not moved back into school is still a scientist and a science teacher, but is now *primarily* an 'academic'. The other changed again to being *primarily* a 'Headteacher'. If this point seems laboured, then we emphasise that when moving into a distinct area of work it is essential to be *inducted* into the new professional identity if one is to be a success in that area of work. This section of the book aims to help with this.

A WORD ABOUT PROFESSIONAL INDUCTION

Of course, something as central as one's professional sense of identity does not change easily. Enrolling on a teacher training course, or accepting a post as a

 14

BOX 2 THE ASSOCIATION FOR SCIENCE EDUCATION

The ASE is the professional association for teachers of science.

The ASE exists to improve the teaching of science:

- by providing an authoritative medium through which the opinions of teachers of science may be expressed on educational matters and
- by affording a means of communication among all persons and bodies of persons concerned with the teaching of science in particular and education in general.

Membership of the Association is now of the order of 20,000, with a broad spread of membership from primary and secondary teachers, to technicians, those involved in Initial Teacher Education, and also includes some 3,500 student members.

The ASE is a publisher. As well as its main journals, *Education in Science*, *Primary Science Review*, *School Science Review* and *Science Teacher Education*, ASE publishes about ten new titles a year.

(Text edited from www.ase.org.uk/htm/thease/history.php,

accessed 6 April 2004)

---

teacher, may make someone a teacher 'by name', but not necessarily 'by nature'. The process of coming to see oneself as a professional science teacher, and understanding what that might mean, does not take place overnight. We see this as a gradual process that takes place over a period of several years. Certainly, we would see the training year and the NQT year as part of this process. This book is primarily directed to 'inductees' into science teaching: those in the process of taking on the new identity of science teacher, and making personal sense of this identity, whether training or already employed as a teacher. We hope that this book will help readers explore what their new professional identity means. We also hope our book will contribute to anyone supervising or mentoring those undergoing this process.

WHAT IS THE SUBJECT MATTER OF SCIENCE?

This is a central issue for science teachers for two main reasons. First, new entrants to science teaching often begin their training by thinking of themselves primarily as biologists, physicists, engineers, doctors, former sales representatives or even students rather than 'scientists'. Professional identities are usually closely linked to a person's previous 'life', as of course this is what is familiar. Time is required to make the mental shift to a new identity. Induction into secondary teaching requires new entrants to adopt the identity 'science teacher', necessitating for many a significant transformation in self-perception as an ongoing process. Thus, one reason for discussing the subject matter of science is to provide material suitable for those making this transition, with the aim of helping smooth the process.

Second, 'school science' does not relate to a real academic discipline. School science is derived from SCIENCE, but based on what science educators and others who designed the NC considered important for children to know. School science and SCIENCE are not the same. This section introduces readers to our perception of school science, as opposed to SCIENCE.

A third reason is that science teachers in most schools are required to teach 'outside their specialism'. This means that although they usually trained as scientists in one broad area of science and then primarily to teach chemistry, physics or biology specifically, there is a requirement for them to train to teach 'science'. This creates a professional dilemma – on one hand, science teachers are regarded and respected for their specialist skills, but, on the other hand, they are expected also to teach as 'experts' throughout the whole science area. Therefore, we try to provide information to help bridge knowledge gaps.

A new entrant's thinking about his/her science is usually well developed and may be to an 'expert' level, perhaps to higher degree standard. However, this is expertise in a narrow discipline within science rather than expertise in science as a whole. Most inductees need to revisit aspects of the broader definition of science that perhaps they met while at school. One important aspect of the narrow nature of most inductees' science subject knowledge is that they share misconceptions about aspects of science with school students: physicists and engineers, for example, may have little understanding about significant aspects of school biology, while many biologists understand relatively few physics concepts. 'School science' requires all science teachers to be able to teach across all disciplines to age 14 and, in many schools, to age 16. Inductees need to subsume their expertise and particular enthusiasm for specific areas of interest into the broader science spectrum.

WHAT COULD SCHOOL SCIENCE BECOME?

Chapters 3 and 4 introduce readers to the wider context of science education. SCIENCE is a major activity in society, contributing significantly to positive changes in lifestyle and life expectancy. Although we cannot predict the major scientific issues of the future, science education can contribute to students' knowledge and understanding of science by providing:

- the confidence to understand (at least some) science knowledge;
- an interest in and the skills needed to research scientific issues;
- an understanding of the processes of SCIENCE, the nature of scientific evidence and the way this is collected, disseminated and evaluated.

Relatively few school students are likely to pursue science-based careers. Science will not be central to the work of the majority, even though the knowledge it produces, and the technologies developed from this knowledge may well be. Hence, it is important to teach not just some science but also to teach *about* science: the place of science in industry, the economy and health-care must be considered. Science lessons can also contribute to the wider aims of education. Developing basic skills such as those of language, co-operative working, problem-solving and data analysis through science will provide school students with transferable skills suitable for any employment field. Citizenship is now a formal part of the secondary curriculum in schools (see Box 5.1), and school science has a full roll to play in the 'new' subject of Citizenship, just as it already makes a major contribution (in terms of informing students about sex, drugs, diet, exercise, etc.) to the now established curriculum area of 'PSHE' – personal, social and health education.

Therefore, in Chapter 3 we take on the task of prompting an inductee to reflect on school science. We offer readers an introduction to how school science as practised in the early twenty-first century in the UK came to be, as well as describing recent moves to ensure progress is made. We invite readers to reflect on their experiences of school science and to consider what they think are important features of a curriculum. We include this chapter to provide a sense of direction – where the school subject has come from and where it might lead. We hope that readers will be stimulated to engage with the debate about the future for our subject.

NEW PERSPECTIVES ON SCHOOL SCIENCE

This perspective on what science education should provide – science for the citizen and not just for the future scientist – is reflected in recent and current developments in the curriculum, discussed in Chapter 4. Inductees will gradually become used to the fact that the ground under a science teacher's feet rarely stands still. Although the NC has not changed in recent years – changes are due – there have been significant curriculum developments offering alternative ways of teaching science. Only by appreciating the background behind new initiatives can the inductee be put in a position to critically reflect on their purposes, methods and potential for contributing to a successful science education.

WHAT DOES A SCIENCE TEACHER'S EXPERT KNOWLEDGE LOOK LIKE?

In Chapter 5 we reflect and comment on what knowledge a science teacher requires in order to be fully proficient in school science. We draw on a range of perspectives, in the anticipation that readers will be encouraged to explore their strengths and weaknesses.

Chapter 2

What is science?

New acquaintance: Oh, so you're a teacher. What do you teach?
Teacher: Children.

INTRODUCTION: WHAT DO SCIENCE TEACHERS TEACH?

The response above appeals to many teachers because it is not possible to teach a subject unless one has someone to teach. Being a school or college science teacher is much more about the students than it is about science. Most new entrants to science teaching are well aware of setting out on a career involving working with children and adolescents – typically hundreds during any week – as a main focus of their work. For many people this aspect of the job is one of the most appealing, either *because* they like working with children, or because they feel they can make a difference to the young people's lives.

Of course, being a science teacher also means spending much of one's working week dealing with aspects of science. This provides a second reason for taking on the job. Many people entering teaching are looking for a career where they use the expertise that their academic (and often industrial) training provides. Sometimes a new teacher will enter the profession primarily because they see teaching as a job that will enable them to continue to work in, and intellectually engage with, their degree subject. Occasionally, such a person sees their work as being substantially *about* science, with the students being something of an incidental feature. This is not a useful mind-set for a teacher, as the learners are always the focus of a teacher's work.

If you are reading this book, then you are at least considering teaching as a possible career. We anticipate that most readers will be in the process of training to become, or be recently qualified as, science teachers. Those in training or in post will have seen and experienced enough to have realised that teaching has to primarily be about the learners and not the subject. If you are only at the stage of *considering* a career in teaching and your motivation for this is, primarily, related to continuing to work in SCIENCE, rather than wanting to teach, then careful reflection about your next step is worthwhile. Here are some points about teaching as a career to consider:

- you will be judged according to how students learn, not how well (you think) you teach;
- you will have little, if any, opportunity to engage with your subject at the level you are familiar with from undergraduate and/or post-graduate studies or professional level work;
- regardless of your degree subject, you will probably be expected to teach across the range of sciences: astronomy, biology, chemistry, earth and environmental sciences, and physics;
- you will be expected to teach a version of SCIENCE that may not be that familiar to you (and may sometimes seem decidedly dubious), that we will label 'school science' (Gilbert *et al.*, 1982).

We will be exploring some of these issues in some detail later in this book, but at this point it is important to point out that we are not trying to put off potential science teachers, as there is, after all, a shortfall of suitable recruits. Rather, we hope to highlight the nature of science teaching and, therefore, the transition required to take on the new professional identity of 'science teacher'.

We state clearly that science teaching is a truly rewarding profession, offering intellectual challenges of the highest order (Taber, 2002a). We need clever, deep thinking science teachers, but the hardest thinking will often be about aspects of pedagogy – how to teach – rather than about the science itself. A good science teacher engages in high-level thinking in planning, executing and evaluating the classroom work. When done well, this thinking is critical and sometimes self-critical, reflective, analytical, creative and informed both by relevant theoretical frameworks and by research evidence. This sounds just like the type of thinking that scientists are trained for: the only differences are that teaching is informed by a different set of perspectives and theoretical frameworks than SCIENCE, and that educational research is necessarily somewhat different from research in the natural sciences.

THE UNITY OF SCIENCE?

Up to this point we have referred to SCIENCE generally, as though it is a unified field of study and practice, although few new entrants to science teaching have degrees in 'SCIENCE' per se. It is much more likely that a new science teacher will hold a degree in a more specialised field such as industrial chemistry, astrophysics, genetics, structural engineering or biomedical sciences. This raises the question of what, if anything, 'SCIENCE' is and whether these various practitioners can be considered to have received any training in common, perhaps something that can be recognised and labelled as 'SCIENCE'.

The most obvious diversity in the academic preparation of new entrants to science teaching is in their subject knowledge. All science graduates know *a lot of science*, but not *the same science*. As many schools (and certainly the requirements for achieving qualified teacher status, QTS) expect science teachers to teach across the sciences at least to age 14 and sometimes to age 16, this has the potential for being a real issue. Many life sciences graduates, for example, may be uncertain about their physics knowledge, while many engineers will have limited knowledge of biology. Few new entrants will have studied aspects of all of the areas of science in the curriculum at degree level and many have not studied at least one of biology, chemistry and physics since the age of 14 or 16.

At first, this may appear to be a major problem in training to be a teacher for the English system because:

- the statutory training standards require teachers to have secure knowledge and understanding of *the subject(s) they are trained to teach* at a standard equivalent to degree level; and
- teachers are trained to teach the NC subject 'science' – see Table 2.1.

Taken literally, this would mean that all trainee science teachers would need degree-level knowledge of the subject called (in the school curriculum) 'science', which could be interpreted as having degree-level knowledge in all areas of SCIENCE that feature in the curriculum. Such an interpretation is clearly not sensible as new entrants to the profession never have degree-level knowledge of all areas.

If this appears intimidating, remember that school science taught up to KS4 is not at a very high level, and any science or engineering graduate *should* be able to learn the 'knowledge' required. The usual interpretation

Table 2.1 Science topics in the National Curriculum

AT	Sc1 Scientific enquiry	Sc2 Life processes and living things	Sc3 Materials and their properties	Sc4 Physical processes
KS1	1 Ideas and evidence in science 2 Investigative skills	1 Life processes 2 Humans and other animals 3 Green plants 4 Variation and classification 5 Living things in their environment	1 Grouping materials 2 Changing materials	1 Electricity 2 Forces and motion 3 Light and sound
KS2	1 Ideas and evidence in science 2 Investigative skills	1 Life processes 2 Humans and other animals 3 Green plants 4 Variation and classification 5 Living things in their environment	1 Grouping and classifying materials 2 Changing materials 3 Separating mixtures of materials	1 Electricity 2 Forces and motion 3 Light and sound 4 The Earth and beyond
KS3	1 Ideas and evidence in science 2 Investigative skills	1 Cells and cell functions 2 Humans as organisms 3 Green plants as organisms	1 Classifying materials 2 Changing materials 3 Patterns of behaviour	1 Electricity and magnetism 2 Forces and motion 3 Light and sound

	Sc1	Sc2	Sc3	Sc4
		4 Variation, classification and inheritance 5 Living things in their environment		4 The Earth and beyond 5 Energy resources and energy transfer
KS4 (single award)	1 Ideas and evidence in science 2 Investigative skills	1 Cell activity 2 Humans as organisms 3 Variation, inheritance and evolution 4 Living things in their environment	1 Classifying materials 2 Changing materials 3 Patterns of behaviour	1 Electricity 2 Waves 3 The Earth and beyond 4 Energy resources and energy transfer 5 Radioactivity
KS4 (double award)	1 Ideas and evidence in science 2 Investigative skills	1 Cell activity 2 Humans as organisms 3 Green plants as organisms 4 Variation, inheritance and evolution 5 Living things in their environment	1 Classifying materials 2 Changing materials 3 Patterns of behaviour	1 Electricity 2 Forces and motion 3 Waves 4 The Earth and beyond 5 Energy resources and energy transfer 6 Radioactivity

is that new entrants' degree-level studies overlap substantially with NC science topics. So, if an electrical engineering graduate's degree included a range of topics such as principles of electricity, circuit design, and so forth, a judgement would be made about whether 50 per cent or more of their degree was in areas linked to the NC. The lack of any mention of, say, green plants, in their degree would not prohibit them from initial teacher training. HEIs also look for the capability of new entrants to learn new material and a flexibility in their thinking indicative of an ability to take on new ideas and communicate these.

This is reassuring for readers planning to enter science teaching with degrees in geology, astrophysics, genetics, chemical engineering, marine ecology and so on. These degree subjects are likely to have provided courses that link to aspects of the NC, even though they are unlikely to have provided material that substantially covers science in the school curriculum.

However, there is something to be explained here: why it is considered *so* important that a new entrant's degree should be in an area of SCIENCE that matches the curriculum, even if it is only directly relevant to a *small proportion* of the topics they will be expected to teach?

WHAT DOES 'SUBJECT KNOWLEDGE AT DEGREE LEVEL' IN SCIENCE MEAN IN TEACHING?

Our view is that this is a genuine and significant issue: the logical consequences of this policy do not seem to have been fully thought through. Few, if any new entrants will have degree-level education in all NC topics and there is hardly time to acquire degree-level knowledge in a year of intensive post-graduate teacher education.

Yet, if we assume that the 'degree-level' stipulation does not apply to the whole science curriculum, but to having learnt some relevant science at degree level, we are then left with an almost meaningless requirement. Perhaps this is just as well, because it allows those involved in selecting and training the teachers to interpret the guidance. The only sensible approach is to assume that the guidance is suggesting there is something common about science (or engineering) education to degree level that equips the graduate to teach school science. There is also a notion that teaching should be regarded as a 'graduate' profession and as such requires that only graduates can enter. To teach science it would seem sensible, therefore, that entrants show evidence of science knowledge gained through a basic degree, even if this does not cover all aspects of the NC. This situation illustrates our point – that SCIENCE and school science are different.

The implicit assumption here is that SCIENCE is considered a recognisable unitary activity *in some sense* and that scientists from different backgrounds share something more fundamental than the subject matter that they study. The school subject labelled 'science' provides students with 'images' of SCIENCE (Millar, 1989c), and should reflect this 'essence of SCIENCE'. These assumptions, if taken together, would support the notion that a graduate scientist would have been educated in this 'essence of SCIENCE' and so have a starting point from which they can communicate this to students.

It may help to consider a different subject area, say history. Two different history graduates may have studied completely different subject matter in their degrees – but they would both have trained in the basics of the discipline of history. These basics would enable them to apply the common conceptual frameworks and technical skills of history to any period and so prepare for teaching the period by suitable background reading.

Clearly, the *logic* of the science curriculum, that is, 'science' as a school subject, is that the same principles apply in science as in history. Botany, industrial chemistry and theoretical physics are considered to be cognate subjects such that degree-level education prepares the graduate in the basic conceptual frameworks and skills of SCIENCE. However, it is our view that this is clearly *not* the case.

Indeed, we would suggest that there is a better analogy with the history graduate employed as a 'humanities' teacher, considered as someone with a 'HUMANITIES degree' and so expected to teach geography and religious studies as well as history, which is not uncommon. If we do not accept that the various sciences should be simply treated as different varieties of SCIENCE, then we need to offer an alternative interpretation of what we think it means to be a 'science specialist' teaching the science curriculum. We offer this framework later in this chapter, but first we give support for our views for readers who may be unconvinced that the notion of 'being a scientist' is inherently problematic.

THE DISCIPLINARY STRUCTURE OF SCIENCE

Let us explore the logic of the government's recruitment requirements for science teachers next from the perspective of a philosophy of science.

According to the well-respected (and physics trained) historian-philosopher of science, Thomas Kuhn, the disciplines of science have limited overlap (Kuhn, 1996). According to Kuhn, SCIENCE comprises a range of distinct traditions that may differ in much more than the topics for study. Kuhn is well known (more so in the social sciences than SCIENCE) for his use of the terms 'paradigm' and 'paradigm-shift'. The latter term is used to describe major transitions, or 'revolutions' in scientific thinking. Later, however, Kuhn preferred to talk of the 'disciplinary matrix' that defined a scientific area (Kuhn, 1977). According to Kuhn's ideas we can think of the SCIENCES as distinct disciplines, each with its own ways of working. The disciplinary matrix has a number of functions: it

- provides the theoretical basis of the sub-branch of science;
- is accepted by all the workers in the field;
- determines what is judged to be the subject of legitimate research in the field;
- determines the procedures, rules and standards that apply in the field.

This means that, in his view, learning to be a biochemist is fundamentally different from learning to be a mechanical engineer or an ecologist.

It is also worth considering Kuhn's initial choice of the term 'paradigm' to characterise scientific disciplines. Kuhn's historical studies led him to the conclusion that scientists were inducted into a scientific discipline largely through learning 'by doing' rather than 'learning about'. Scientific training was largely achieved by working through example problems set by those already inducted into the discipline, rather than just being directly taught. This activity led Kuhn to the initial meaning for 'paradigm'. Kuhn suggests the nature of the set problems changes as the scientist-in-training proceeds from first degree to graduate work.

In other words, training as a scientist is *more than* an undergraduate education. The professional bodies awarding designations such as 'Chartered Chemist' and 'Member of the Institute of Electrical Engineers' recognise this. We would go further, suggesting that most undergraduate courses offer limited opportunities to undertake the type of professional preparation that Kuhn sees as central to being inducted into a scientific discipline: usually, a first degree in science involves much more learning *about* the knowledge produced by a SCIENCE than learning to *do* the SCIENCE.

Some new entrants to science teaching have higher degrees and research experience in their discipline, and so can be considered to be fully inducted scientists in Kuhn's sense. However, most science teachers have little or no genuine research experience. This would suggest that science teachers, en bloc, are a disparate group of individuals with incomplete training in a range of different disciplines.

SCIENTIFIC ENQUIRY: SC1 AS THE COMMON ELEMENT IN SCIENCE AND SCHOOL SCIENCE

Sc1: scientific enquiry in the National Curriculum

A cursory glance at the science NC (summarised in Table 2.1) provides a clue as to how science as a school subject is framed within the English curriculum. This is the existence of 'Sc1' the distinct theme of *scientific enquiry*.

Sc1 runs alongside Sc2–4 through the four Key Stages of the school curriculum: in other words, it is a constant major feature of science education throughout compulsory schooling. Students should experience the two identified separate aspects of scientific enquiry, namely, 'ideas and evidence' and 'investigative skills' on a regular basis throughout this time. Sc1 is the part of the curriculum intended to reflect the generic processes of SCIENCE. According to the documentation (DfEE, 1999, various pages), scientific enquiry is to be taught in the context of content material taken from Sc2, Sc3 and Sc4: 'Teaching should ensure that scientific enquiry is taught through contexts taken from the sections on life processes and living things, materials and their properties and physical processes.' In practice, Sc1 neither entirely solves our problem of identifying the common elements of the sciences, nor has it been successful in its own terms.

First, in terms of looking for commonality, it is often said that what different sciences have in common is not their content (i.e. subject matter), but their adherence to the scientific method. However, philosophers and sociologists of science who have studied how SCIENCE proceeds have suggested that there is no such thing as a unitary scientific method. Not all commentators would go as far as Paul Feyerabend in his *Against Method* (1988) and imply that 'in SCIENCE, anything goes'. Yet, as Kuhn's notion of the disciplinary matrix highlights, the actual techniques of data collection and analysis vary considerably in different sciences. We explore the issue of content versus process in the context of school science in Chapter 3.

Different scientists do not only use different scientific apparatus but they may actually have different approaches to what can be considered appropriate and valid methodology. For example, the process of SCIENCE is commonly considered to involve 'experiments', but there are sciences (such as anthropology) where data collection is not usually based upon an experimental paradigm. In some sciences, it is normal and necessary for data to be analysed in terms of inferential statistics in order to be able to draw conclusions, but in other sciences the data does not need to be considered in this way.

In Sc1 students are taught 'investigative skills' that supposedly reflect the processes of SCIENCE, when we contend that those processes cannot be simply defined or characterised in a way which can well match the investigations of scientists in general, be they particle physicists, synthetic organic chemists, population biologists, field geologists or radio-astronomers.

Sc1, then, offers a view of the processes of SCIENCE that must be considered as something other that a prescription for 'the' scientific method. Rather, the view of SCIENCE presented in Sc 1 is a *model* of scientific process.

The notion of a curriculum model

Models of various types are, of course, familiar to many scientists, as simplifications and representations of complex systems and as tools for thinking and communicating ideas.

Models are used in education for many of the same reasons as in SCIENCE (Gilbert and Boulter, 2000). Consider a topic such as photosynthesis. This is an important scientific topic, and one which is potentially very complicated. Secondary age students should learn about photosynthesis, as an understanding of this leads to understanding about the interdependence of living things and the essential role that plants play in maintaining our environment and providing food. If we believe science education should help students understand the world in which they live, then it is difficult to think of many more important topics. Yet, teaching the details of photosynthesis to 11–16-year-olds is neither possible nor desirable. Students should be taught some key aspects of the topic to give them a basic understanding; for example, they need to understand that photosynthesis depends upon sunlight, but it is not important for them to know the mechanisms by which energy is absorbed and transferred in the photosynthetic system.

The same principle applies to other topics: we cannot teach the full technical details of cutting-edge SCIENCE. The curriculum necessarily, and quite rightly, presents a simplification of the scientific knowledge available. In some cases, we might genuinely consider the simplification to be simply the selection of what are judged the most essential points and the omission of other less central details. Yet, in practice, we know it is not that simple. The 'meaning' of any scientific concept derives from its relationship to other scientific concepts. It is also the case that many important scientific concepts are inherently subtle, abstract and complex. Light is an example. At KS3 light tends to be considered as a wave phenomenon and its particle-like properties are largely ignored. This is a simplification, but not in the sense that the particle-like nature of light is less central than its wave-like properties. Rather, a judgement is made that the scientific concept of light needs to be simplified for pedagogic reasons, to give a version that students of secondary age may be able to make sense of (cf. Shayer and Adey, 1981). Wave-particle duality *is* taught at 'sixth form' level.

The 'version' of the light concept presented to 11–16-year-olds is *a curriculum model*, of scientific knowledge matched against what experienced educators reasonably believe students of this age may make sense of. A better appreciation of the nature of light and its behaviour would be given by using Quantum Electrodynamics (QED) as the basic model: but that would not make sense to many students (or some teachers).

The curriculum model of light used in school science may not match current best SCIENCE, but is probably close to the historical model regarded as 'best SCIENCE' for some considerable time. We often find that curriculum models are quite similar to historical scientific models (Justi and Gilbert, 2000). In addition, the curriculum model of light used in secondary science may match quite well some science teachers' mental models.

Conceptual fossils in the curriculum

The presence of some curriculum models may reflect science teachers' 'comfort zones' rather than being good science teaching. Models that once had scientific validity and utility, but which are now little more than historical curiosities, can easily be passed down through generations of science teachers, aided by school textbook authors and sometimes examination questions. This process of conceptual 'fossilisation' was pointed out by the French science teacher and philosopher, Gaston Bachelard (1968), as long ago as 1940. A curriculum model may be a good simplification at the level being taught because it is easy to learn. However, these judgements are being made by those already well familiar with the models, namely the teachers, not the students, and being 'easy to learn' should not in itself justify the presence of a dubious model in the curriculum (Taber, 2003a).

Figure 2.1 shows a common diagram, as drawn by a post-16 student and meant to show ionic bonding. In fact, the diagram shows something quite different: the transfer of an electron from an atom of a metallic element to an atom of a non-metallic element. The figure shows *ion formation*, which is a different phenomenon from *ionic bonding*. The student should be forgiven for this mistake, as many school textbooks and teachers identify diagrams such as this with ionic bonding (Taber, 1994) and teach the topic accordingly. Students have even been asked to reproduce such figures in GCSE examinations as 'showing ionic bonding'.

Figure 2.1 does not represent ionic bonding or a likely process by which ionic bonding might come about. Ionic bonding is a lattice phenomenon, involving large numbers of ions, not a single cation and a single anion. If secondary students prepared sodium chloride (the usual 'model' ionic substance), they would most likely use solutions of hydrochloric acid and sodium hydroxide, in which the ions are already

Ionic bonding

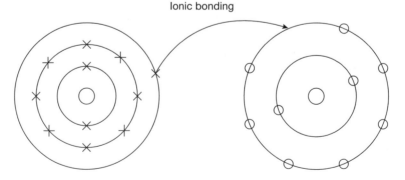

Figure 2.1 *Ionic bonding?*

present, and no electron transfer is needed. Even if sodium were burnt directly in chlorine (not by the students!) the reaction would involve sodium metal and molecular chlorine gas, not discrete atoms. Therefore, Figure 2.1, is not a good representation for ionic bonding, yet these diagrams are ubiquitous.

The NC at KS4 requires that students 'should be taught: that chemical bonding can be explained in terms of *the transfer* or sharing of electrons; [and] how ions are formed when atoms gain or lose electrons and how giant ionic lattices are held together by the attraction between oppositely charged ions' (DfES/QCA, 1999). In scientific terms, chemical bonding is about the forces holding chemical structures together and can *not* be explained in terms of the transfer of electrons; but clearly, this has become established as a curriculum model.

The principle of the optimum level of simplification

So this is what we mean by a curriculum model: a formalisation of some aspect of science, simplified for pedagogic purposes to make the science accessible to the learner. We are perfectly happy with *the principle* of curriculum models, although (as in the example of ionic bonding) we think they should be explored critically to ensure they hold both scientific validity and pedagogic utility.

Simplification, then, is a necessary, but not sufficient, condition for an effective curriculum model. We would argue that curriculum models should aim for the optimum level of simplification (Taber, 2000a): simplifying enough for students to make good sense of them, yet providing a good basis for progression to more sophisticated levels of understanding.

A curriculum model of scientific enquiry

We would characterise Sc1, then, as providing a curriculum model of scientific enquiry. In effect, Sc1 has two distinct parts, both appearing under the heading of 'scientific enquiry' and both have proved problematic in teaching. In both cases, the difficulties are related to aspects of curriculum assessment.

The first aspect, 'ideas and evidence in science', has not until recently been explicitly assessed, and there is a perception that teaching of it has been 'patchy'. The second aspect, 'investigative skills', has been formerly assessed at GCSE. The assessment has been divided into four features:

planning an investigation, collecting data, analysis of collected data and evaluation of the investigation. These four aspects of investigation represent a simplified research cycle and, as such, may be considered a curriculum model of how to undertake a scientific investigation: that is, a school curriculum version of scientific method.

In principle, this seems sound. A student has to plan what to investigate and how to go about it. S/he then needs techniques to ensure collection of valid data. The data is then analysed to see what, if any, conclusions can be drawn. Finally, the student then suggests how the investigation could be improved.

In practice, Sc1 has become a formulaic exercise. First, each component is assessed separately according to a set of hierarchical criteria. These mean that students cannot access higher scores unless the criteria for the lesser categories are clearly met (see Chapter 10). Such a scheme is probably essential to ensure objectivity of marking, as the investigations are marked by the teacher in school rather than independently by the examination board.

The logic for this development was to assess students' practical skills, which would be demonstrated under the teacher's supervision and partly to offer schools flexibility to devise their own 'assessed practical exercises'. Both of these worthy intentions became somewhat distorted in practice. As examination boards required evidence to ensure all schools were marking to the same standard, it soon became clear that what was to be assessed was not the student's investigation skills per se, but their *written report* of the investigation.

Schools could devise their own investigations to meet local needs and preferences within a general, nationally agreed framework. The assumption was also that science teachers could carry out the investigative work to fit in with their SoWs. At the end of the GCSE course teachers would select examples of students' practical work to assess against the criteria. A sample of work would be sent for 'moderation', to ensure comparability of marking standards between schools. As often happens with well-meaning ideas, the reality was that a 'lowest-common-denominator' mentality soon descended upon the whole process.

Teachers soon found, for example, that examination boards had specific ways of interpreting the criteria, so leading to the realisation that certain types of practical work did not permit students to obtain the highest marks in some of the four assessed areas. Genuinely open-ended investigations also tended to limit the ability of most students to meet all the criteria, at least in the amounts of class time available. Over a period of time, what has happened in most schools is that a small set of standard 'investigations' have evolved that are highly constrained, but are known to enable most students to meet the criteria within the time constraints. 'Suitable' investigations included: investigating the effect on resistance of different lengths *or* different thicknesses of wire and investigating the effect of temperature *or* concentration of acid on the rate of reaction between zinc and hydrochloric acid. Schools found the minimum number of investigations needed to allow students to practise and be given feedback, and then to provide the essential evidence for the assessment.

This approach leads to a mechanical, formulaic approach to Sc1 investigations, which was not intended by the curriculum developers. In SCIENCE, genuine investigations may be open ended and speculative – for example Rutherford and his co-workers carried out some 'silly' experiments at the Cavendish Laboratory, just to see what might happen (Cathcart, 2004). In school science, an Sc1 investigation nearly always leads to an outcome that students already know or could find out from a textbook. Schools programme 'investigations', often referred to as 'coursework', into their schemes of work as stand-alone episodes, rather than allowing an investigation to flow naturally from a science topic being taught at any time. This contradicts both the original intention of assessing students' skills and the way that empirical and theoretical work are mutually reinforcing and sustaining in SCIENCE. However, given an assessment system where most students can be trained to obtain good Sc1 marks by practice and focusing on meeting the criteria, schools feel obliged to do what they think is necessary to get students the best scores.

Even more disappointingly, the range of science practical work in some schools seems to have been reduced to allow more time for 'theory'; after all, Sc2–4 provides a great deal of material to be 'covered', while the only practical work formally assessed are the Sc1 investigations.

This does not mean that the curriculum model for scientific investigations is poor *as a model*. But the way the model is interpreted and operationalised in many schools hardly provides students with insight into the range of empirical work undertaken in SCIENCE. For example, within the National Strategy for Science at KS3 (discussed in Chapter 3), teachers are encouraged to take a wider view of what investigative work may mean in school science (see Box 2.1). The Strategy also highlights how Sc1 *should* permeate teaching about all science topics:

BOX 2.1 SCIENTIFIC ENQUIRY IN THE KS3 NATIONAL STRATEGY

There are many different types of scientific enquiry such as:

- pattern seeking – for example in surveys or correlations;
- using first-hand and secondary sources of information;
- identification and classification;
- using and evaluating a technique or technological application;
- fair tests involving the control of variables;
- using experiments models and analogies to explore an explanation, hypothesis or theory.

(DfES, 2002a: 12)

Scientific enquiry generally links practical experience with key scientific ideas. In *the most effective practice*, the principles of scientific enquiry are not left to special 'investigative science' lessons. They are integrated into most lessons, even those that involve little or no practical work. Teachers capitalise on chances in any lesson to encourage pupils to reflect, however briefly, on the evidence that supports scientific interpretations.

(DfES, 2002: 11, *emphasis* added)

That the government believes there is a need to make these points as part of an expensive and extensive programme of curriculum advice and in-service training (see Chapter 4), highlights that in practice this is not what usually happens. With limited attention to the 'ideas and evidence' thread of Sc1 in many schools, this attainment target (AT) is not currently acting as a sound basis for providing students with a vision of the common ground in SCIENCE.

RECAP: WHAT SCIENCE AND SCHOOL SCIENCE DO NOT, AND MIGHT HAVE, IN COMMON

First, the 'do not'

This chapter has explored the question 'What is SCIENCE?' because unless we have some idea of what SCIENCE is, it becomes difficult to see how this can be reflected in the school curriculum.

We have suggested that SCIENCE cannot be characterised in terms of subject matter, or even how this subject matter is explored. Within SCIENCE, there is a range of distinct sciences. We do not suggest that the subject matter of SCIENCE 'naturally' breaks down into different subjects; the border between physics and chemistry, for example, is permeable and rather fuzzy in places (Taber, 2003b). However, as SCIENCE developed and became too vast for any one person to master, it transformed into specialist areas that in time developed their own traditions and ways of working. These discrete disciplines evolved over time due to a range of institutional and political factors and the historical contingency of the influence of scientific 'greats' (Aikenhead, 2003).

School science does not, and should not, fully reflect the disparate nature of mature SCIENCE. Students who are learning science as part of a wide range of curriculum subjects need to be presented with something that will appear to have a certain degree of coherence and unity as a school subject. That is not to say that the manifold nature of the sciences should be invisible, but it should not overcomplicate the subject for the students. Curriculum science needs to provide a simplification of the disciplinary structure of SCIENCE for the benefit of learners.

The government training standards seem to imply that all science teachers should have degree-level knowledge in all the areas of the science curriculum they will teach, which means across biology, chemistry and physics for most science teachers. We have pointed out that this is not a feasible policy and that, in practice, science teachers have degree-level education that substantially matches a subset of curriculum topics. This inevitably means that individual science teachers only have a strong background knowledge in some parts of the science they will teach, something that can lead to aspects of 'teacher's science' being limited and including alternative conceptions (see Chapter 5).

We suggest that the profession is able to respond to this, *provided that* science departments have teachers with complementary science knowledge that allow staff to be timetabled where they are most confident and able to support each other in developing scientific knowledge and understanding. We point out in Chapter 5 that this might not be the case in many schools.

A naive view would hold that subject matter does not define SCIENCE, rather the process does: that is, the scientific method. Yet Kuhn shows us that scientific methodology is, itself, specialised within disciplines. Moreover, the model of scientific investigations presented in the curriculum is not only limited to the type of work undertaken in a subset of SCIENCE disciplines, but has become a formulaic exercise completed with a minimum of creativity and in the minimum curriculum time needed to secure students maximum marks in their GCSE Science coursework.

This might disappoint the new science graduate hoping to pass on their enthusiasm for SCIENCE through teaching. We believe there are some real issues here that can lead to science being seen as a content-heavy subject with little scope for imagination or time for reflection. However, we are optimistic, because we have seen how good teachers and trainees manage to produce exciting and valuable lessons, even working within this curriculum 'straitjacket'. The situation does need to change and may be beginning to change in some areas (see Chapter 3). Part of our message in this book is that a role for the *professional* science teacher is to work to bring about better science teaching in schools. At this point, however, we wish to end the chapter on a positive note, by suggesting what might form the commonality between SCIENCE and school science, that can act as a platform for moving science teaching forward.

The 'might have': an alternative basis for common ground in SCIENCE and science education

If subject matter and method cannot be considered to be shared across SCIENCE and therefore offer no basis for common ground between SCIENCE and curriculum science, then we need to look elsewhere for that common ground. This is more at the level of *values and attitudes* than in terms of specific ideas or skills. We suggest that there may be no single scientific method, but we may be able to agree on what can be called 'a scientific attitude'.

If professional science teachers are to exert their professionalism by wresting more control on what school science is and becomes, then we will need to work through organisations, such as the ASE. The ASE already has policies, developed over many years, on a range of topics including one on values and science education. Box 2.2 presents the values suggested by the ASE and further information is available from the website (www.ase.org.uk, accessed 1 August 2004). Although this may not be the precise list we, or the reader, might have compiled, this is a good starting point for considering what is fundamental about SCIENCE and so should be central in science education.

BOX 2.2 VALUES AND SCIENCE EDUCATION

The Association believes that values which guide scientists' conduct include expectations:

- to be thorough in all operations, including observation, calculation and reporting;
- to be intellectually honest, e.g. refraining from exaggeration, not plagiarizing;
- to be open-minded, e.g. willing to look for and consider new evidence, facts and theories;
- to suspend judgement rather than make snap judgements;
- to be self-critical and to encourage others to criticise one's work.

Science education can contribute to learners' development as whole people – morally, intellectually, aesthetically, culturally, emotionally. Education in science can help learners to develop:

- an enthusiastic interest in and a constructively critical attitude towards scientific values, ways of working and ways of seeing the world;
- curiosity in and a responsible attitude towards the natural and physical world;
- a considered appreciation of science as a creative, human activity which both influences people's lives and views of the world and is itself subject to potent social and cultural influences;
- an ability to connect and apply scientific knowledge wisely in response to practical problems and issues facing individuals and communities;
- critical thinking about phenomena, weighing of scientific evidence and systematic use of skills.

(From the ASE's policy statement on *Value and Science Education*. From the ASE website, www.ase.org.uk, accessed 6 April 2004)

We also suggest that the aspect of Sc1 labelled as 'ideas and evidence' has much potential for teaching about common ground in SCIENCE. All sciences are based around theories developed through empirical investigations, leading to model building used to explain and predict. Although the empirical work does not always match the 'controlling variables' model currently taught in Sc1 investigations, the interplay between evidence and theory and the development of models and theories for use in explanations is surely central to all SCIENCE. If Sc1 'investigations' cannot reflect this, then the use of historical and contemporary cases as contexts for exploring the ideas and evidence strand certainly has the potential to give students a taste of what SCIENCE is really about.

THINKING ABOUT PRACTICE

Here are some questions for reflection on the material in this chapter:

- What exactly is SCIENCE?
- Write a list of the first 20 famous scientists you think of. Look at the list and consider: how many are women?; how many are from outside Europe or the US?; how many made their discoveries in the last 30 years? What does this tell you about your view of what SCIENCE is? How might this need to change in the process of becoming a professional science teacher? (adapted from Reiss, 2002a)
- What is common to all scientific disciplines?
- Is there such a thing as *the* scientific method?
- Where does your science knowledge overlap with the NC summary in Table 2.1? What additional knowledge would you need in order to teach school science?

Chapter 3

What could school science become?

Change is not made without inconvenience, even from worse to better.
(Samuel Johnson)

INTRODUCTION

In this chapter we reflect on how school science could or should look in the twenty-first century. Initially, the chapter explores how school science achieved its current state. Following this, we introduce a new approach to the science curriculum and examine what remains to be included.

An apology, and a promise

We begin with an apology and a promise. For those who picked up this book assuming that teaching school science is simply teaching SCIENCE in a school or college context the previous chapter may have seemed bewildering if not depressing. We suggested not only that school science is not SCIENCE, but that SCIENCE is difficult to characterise in any straightforward way that allows simple reflection in the school curriculum. We apologise if starting a book in this way seems rather discourteous to our readers – many of whom are hoping for inspiration and support as they consider or enter a new stage of their careers.

We apologise for any discomfort caused, but not for chapter content. Science teachers should acknowledge the way school science deviates from SCIENCE if they are to individually or collectively work towards more valuable and authentic science education. Our belief is that in this case, *things have to get worse before they can get better*. This is because if science teachers ignore the problems with school science, they will not be provoked into improving practice, something that is central to being a professional.

The promise is that this chapter should make more comfortable reading, for here we explore not just what science education currently is, but what it could become.

A frame for making sense of school science

We have argued that teachers teach some alternative entity, something that has been socially constructed, that might be labelled 'curriculum science'. By socially constructed we mean that what happens in schools is not part of some inevitable natural order or the result of divine intervention, but has evolved through processes that might best be labelled as 'socio-historical'. In other words, the curriculum is the outcome of a political process. This, in itself, is not a good or bad thing, just an observation that schooling is regulated and controlled through various processes that are 'political' – in the sense that they are determined and influenced by groups holding various forms of power.

We point out that SCIENCE is similarly socially constructed; even something as fundamental as the number of chemical elements is determined this way. This does not mean that the identification and characterisation of elements is arbitrary, but just that the processes by which results are reported, accepted and regulated are through groups with power, such as journal editors.

In SCIENCE, power may be wielded democratically, fairly and with due consideration for evidence and logic – but it is nonetheless a political process. We hope that today science is guided by values we would aspire to, but this does not make it value-free and totally objective.

As in SCIENCE, so with education: policy is decided by groups with power, and influenced by groups with the resources and determination to have their say. Unlike in SCIENCE, nature is not available as 'final arbiter' to moderate the process. Election of a racist party to government would most likely result in a racist education policy. Although such a policy would not be consistent with our values, it could not be shown to be 'wrong' in terms of scientific evidence; indeed, it may well be the 'right' policy to further the interests of the group in power. If this seems far-fetched, then bear in mind that pressure groups in some parts of the US have achieved policies whereby so-called 'creation science' must be taught in schools and have parity of time with teaching about evolutionary ideas. Similar pressure groups are trying to enforce comparable policies in other countries that we might imagine to be more enlightened in such matters, such as Australia.

The political nature of human affairs and the socially constructed nature of the curriculum are not in themselves problems to be solved. Rather, they are processes to be understood and engaged with if we wish to ensure that *our* interest groups have influence and *our* value systems inform policy.

Before we can start exerting pressure, however, we need a good understanding of what science education is and what we would actually like science education to be. This chapter asks the questions: What *could* school science look like, and what *should* school science look like? As part of the process, readers are invited to reflect on their own experiences of school science and to consider what they themselves think is appropriate for the future.

Two tensions – 'process' versus 'content' and 'generalist' versus 'specialist'

The two most significant tensions within UK school science have been *process* versus *content* and *generalist* versus *specialist*. The process/content tension lies between teaching 'the scientific method', that is, science as a 'process' and teaching scientific knowledge or content, called a 'content-led' approach. 'Process-oriented' school science focuses on how scientists 'do science'. The aim is to encourage children to think and be trained as scientists work in practice. 'Content-oriented' school science places knowledge about science in terms of factual information as most significant, on the basis that any experiment is meaningless (or, at best, its value is greatly reduced) without the relevant background knowledge. Scientific knowledge also provides facts about the world, universe, how objects work and so on. UK school science originated in the nineteenth century as a process-based subject, but acquisition of factual knowledge became increasingly important during its first 50 years. Although it may seem obvious that both content and process are desirable, the balance between them shifted position during the latter years of the twentieth century.

The contrast between providing 'science for all' and the level of science required for practising scientists is at the heart of the generalist/specialist tension. By 'science for all' we mean school science sufficient for the needs of people living in a society in which science contributes significantly to improvements and complexities in health, lifestyle and well-being. By 'specialist' science, we mean the school science required to maintain a steady flow of well-trained scientists such that the UK's science-based industry can maintain a high level of productivity and international competitiveness. As school science provides the starting point of children's scientific education, there is a perceived need for this to meet the requirements for, and to encourage, careers in SCIENCE. At the same time, schools are also expected to meet the science knowledge needs of those who do not intend to pursue a scientific career. This has created conflicts in terms of curriculum appropriateness, resources, time and assessment.

We will explore briefly the origins of these tensions to help provide background for the situation today.

THE PROCESS V. CONTENT TENSION

The origins of school science lie in process skills

Process-oriented school science, that is, the process of *doing* science, is often referred to as learning 'scientific method'. In practice, it has come to mean learning a set of skills, such as observing, classifying, inferring and hypothesising (see Chapter 2). This tradition provided the basis for UK science education and has retained a significant influence throughout the twentieth century. School science has its origins in the hectic scientific activity of the 1830s, when SCIENCE was the preserve of professional scientists (Jenkins, 1989). As new discoveries were made, new 'sciences' developed, causing scientists to express concern that the subject was being split up into different disciplines, thus lessening SCIENCE's impact on society. In an attempt to avoid this, scientists realised that aspects such as practical activity, observing, classifying and hypothesising were the same regardless of discipline, and so could be brought together as one subject. Thomas Huxley formalised these ideas in 1854, in suggesting that SCIENCE was 'trained and organized common sense' and necessary in a 'gentleman's education', and so should be taught in schools. School science was established from about this time. School laboratories were built, practical texts were written and patterns such as 'test, observation, inference' developed. In 1888, the chemist Henry Armstrong published 'How to teach science', an influential report explaining his 'discovery learning' or 'heuristic' approach. This formalised the early instincts of the 1830s into a teaching strategy. Armstrong's influence was retained well into the twentieth century, losing ground between 1920 and 1950 but resurfacing in the 1960s' Nuffield Curriculum Projects in the UK and other courses elsewhere. These advocated teaching biology, chemistry and physics mainly through students carrying out experiments designed to introduce key concepts in science. Students would be led through 'discoveries' in similar ways to the original work done by scientists.

By the mid-1980s, immediately preceding the introduction of the National Curriculum for Science (in December 1988), courses such as *Warwick Process Science* (Screen, 1986), *Science in Process* (ILEA, 1987) and *Nuffield 11–13* (1986a, b) were popular. These featured training children to develop 'process-skills' such as hypothesising, classifying, observing, predicting, controlling variables, evaluating and inferring. Science knowledge was thought to be irrelevant, partly because these skills were considered transferable and would outlast factual knowledge; and because scientific information was deemed readily available from other sources. GCSE courses became dominated by assessments of these skills, following a pattern not dissimilar to that established in the 1800s. Schools commonly used such courses as lower secondary material in preparation for GCSE.

Criticism of process-oriented school science

Criticism began early in the twentieth century when the impossibility of school science lessons mirroring discoveries such as the basics of atomic structure and quantum theory, let alone those made in the past by scientists such as Newton, was realised. Next, in 1917, came scientists' desire to preserve school science as deserving special attention in education as a subject of extreme importance to society. Huxley's claim that SCIENCE was no more than 'organized common sense' almost succeeded in destroying this position, as it could be argued that there was, therefore, nothing special about SCIENCE. Scientists started to discredit the discovery learning philosophy, replacing this with content knowledge.

More recent criticism began in the 1970s and 1980s. For example, Robin Millar raised three main objections to process-oriented science (Millar, 1989b):

1 that scientists and philosophers have not reached general agreement about what 'scientific method' is;

2 that the processes listed as being part of SCIENCE are not special to SCIENCE, but can be found in other disciplines;

3 that there is no evidence the 'scientific skills' can be taught such that children or adults improve their performance.

The development of constructivist theories of learning (see Chapter 4) contributed to discovery learning being discredited because children's prior knowledge and experiences mean there is no possibility for them to 'discover' as intended. Even basic observations that seem so clear to a trained scientist are not 'seen' in the same way by children (Driver, 1983). Also, the claim that science skills are transferable to other areas of life has not been sustained. Children at school tend to compartmentalise lesson material presented to them. There is no real encouragement or perceived need to use skills learned in science in other subjects and vice versa. Finally, the lack of support from content knowledge means that the reasons why practical experiments are undertaken may not be made clear to students. The lack of background information might, in itself, lead to poor results and understanding.

Content-led science developed from early twentieth-century conditions

The drive towards making school science content-based took root in the early twentieth century. Several positive factors came into play, besides the early criticisms of process-based school science described above. First, the world was at war – chemical warfare and new weapons were being used against allied armies completely ignorant of basic scientific facts. This made it very difficult, for example, for commanders to counter the effects of gas attacks at a practical level and for treatments to be effective. A second factor was reflection on the Armstrong–Huxley conception of school science by the British Association (BA) in 1917. The BA reflected (1918) on the paradox that SCIENCE and science teaching could not be regarded as more 'special' than any other subject if all it comprised was a set of 'common sense' skills applicable to all subjects. Science in schools had to be based on something deeper and more significant to retain its place in the curriculum. It was suggested that 'doing science' could not be separated from the content – scientific method was recognised to only be meaningful in relation to subject knowledge. And third, due as much, perhaps, to the dehumanisation and changes in society created by the war as to the BA report, science was subjected to a shift in philosophical approach. For the first time, school science was to be taught for the benefit of the learner – intellectual development of individuals and knowledge about science in relation to life and the everyday world became a guiding light. As a result, a move towards 'general science' became irresistible.

Initially, the subject was not popular, reaching only 23 per cent of secondary school children by 1942 (Jenkins, 1989). At the same time general science was being proposed, biology was developing as a school subject in its own right, being taught in almost all schools in some form from the 1930s onwards. Reasons for the popularity of biology relative to general science include the range of potential careers involving biological knowledge and the wide range of subjects included under the biology umbrella, such as health, nutrition, anatomy, rural science and social policy.

During this period, the development of content-based science courses varied between types of secondary school. UK 11-year-olds in the state-funded system were selected to attend different, usually single-sex, types of school using results of an IQ-based test called the '11-plus'. Grammar schools took children with the top 10–20 per cent best results, while the remainder went to secondary modern or technical schools. General science featuring physics and chemistry tended to be taught in boys' grammar schools, while biology and other science- (usually biology-) based courses were taught in the other types of school. Thus, biology became regarded as the science for 'less able' students. Content-based curricula developed steadily in all science areas, but grammar schools, which were at that time the main providers of future scientists, led the way with new projects and assessment strategies.

In the late 1960s and through the 1970s, the practice of selecting children for specific schools on an IQ-test basis fell into disrepute. Comprehensive schools for all children replaced almost all selective schools, giving the opportunity and necessity for more consistency to develop science curricula. However, grammar school science curricula continued to dominate, even though this had become highly loaded with content. The overloaded nature of the curriculum was one argument used to justify the Nuffield projects of the 1960s, as these were relatively content-free, being based on the old discovery learning principles.

As the late twentieth century progressed, the science curriculum largely retained its grammar school 'feel'. Content centred on biology, chemistry and physics, with assessment being adjusted to meet students' differing academic needs. Hence, in the mid-1980s CSE and O level science courses were taught side-by-side – very similar in content, but varying in academic demand. The original Science National Curriculum document (DES, 1989) proposed no less than 22 attainment targets, including two about practical work. Each target specified content to be taught between the ages of 5 and 16. At this point, the content-led curriculum had reached its peak.

Criticism of content-led school science

Content-led curricula have been subjected to various criticisms. For example, Jerry Wellington (1989) raises these points:

- a content-led science curriculum is based on ideas that are abstract and can only be learned at best by rote;
- scientific knowledge can be accessed easily using ICT (and today by extension, the world wide web);
- scientific facts date quickly, so should not form the basis for a scientific education;
- transferable skills are more useful to children than knowledge.

There is supporting evidence for these. First, the content-led curriculum developed during the twentieth century is regarded as inappropriate for the vast majority of children. In the UK's case, content developed in grammar schools for those perceived as most suited to studying science at university. Second, the overloading of a curriculum with content knowledge reduces children's opportunities for learning science in a meaningful way, while time pressure to cover the curriculum for terminal assessment means that rote learning of information is often all that is possible realistically. The failure of the curriculum to allow children time to assimilate knowledge before moving to a new topic leads to their dissatisfaction and frustration. Third, the retention of content based on scientific facts established in the distant past has contributed to the subject being perceived as irrelevant for those living in today's society. Scientific progress has continued, but new discoveries find little or no place in a school science course, as completion of the curriculum and assessment priorities dictate. Applications of scientific knowledge in school science courses are few. School science, in contrast to newer, trendy subjects, seems like a rather fussy, pedantic male uncle whose attitudes, dress sense and language are firmly rooted in the nineeenth century.

The process/content tension

The balance in school science between the practical skills involved in doing science and the factual knowledge required to understand the subject has been disputed throughout most of the twentieth century. The tension has been resolved first in one direction and then the other. The first National Curriculum document (DES, 1989) sought a compromise that has been maintained – that practical 'skills' should be an important and assessed part of a heavily content-dominated curriculum, acknowledging the place of both aspects.

However, the criticisms of both process- and content-based science courses are significant. Dissatisfaction with the current situation is widespread among teachers and, not least, children, who are choosing not to study the subject beyond 16. A twenty-first-century curriculum requires a rethink of the position on both of these aspects.

THE GENERALIST V. SPECIALIST TENSION

The rise of 'science for all'

Science became established in the school curriculum, but its uptake remained patchy, partly because provision was divided by school type (and usually also by gender), but also because children were able to choose science as a subject for study. Grammar school girls, for example, could normally access biology, botany and zoology courses at most, as physics and chemistry were available at boys' schools. Those attending technical and secondary modern schools were offered a variety of non-examination (and therefore low-status), mainly biology-based courses such as 'science in the home' and 'science and health'. The prestigious exams required for university entrance, called the School Certificate (taken at age 16) and Higher School Certificate (taken at age 18), and from 1951 'Ordinary' (O) and 'Advanced' (A) levels, were available only to grammar school and exceptionally the 'most able' secondary modern school students. Sixteen- to eighteen-year-olds were forced to decide between being 'arts' or 'science' based, as there was no possibility of taking a combination of higher level courses.

47

By the 1960s government was becoming increasingly aware of the uneven distribution of educational opportunities for 13–16-year-olds. The introduction of comprehensive schools in which teachers and children from several different schools and traditions all met brought these matters into sharp focus. In the case of science, large-scale curriculum reform began, working towards providing courses that could be available to all children. The notion of combining sciences together in an 'integrated' course was developed through the Schools Council Integrated Science Project (SCISP, 1974), which laid the foundation for 'balanced science' which came later. Certificate of Secondary Education (CSE) courses were developed for those not taking O level. Some CSE courses were 'mode 3', which meant that teachers could contribute to course design and content suitable for their local context. However, at this stage, although science was now consistently available, children were still able to choose – some elected not to study any science beyond the age of 14; others chose one or two science subjects and some followed three separate courses leading to examinations in biology, chemistry and physics. Gender bias was prevalent, with mainly boys taking physics, girls picking biology, and chemistry being more evenly divided. Although science provision had improved, the outcomes for 16-year-olds still varied.

The movement towards providing consistent science for everyone, called 'science for all', grew through the 1970s. Several events occurred that helped build momentum. Prime Minister James Callaghan initiated a wide-ranging debate about education in his 1976 Ruskin College speech. In 1979 the ASE published *Alternatives for Science Education* (ASE, 1979) which set out the need for children to have the same entitlement to science. HMI (Her Majesty's Inspectors of Schools) introduced the term 'balanced science' in their 1979 secondary school survey (DES, 1979). Through the early 1980s, the adoption of 'broad and balanced' science for all children aged 5–16 became an established idea, backed finally by the 1985 DES report 'Better Schools'. The 1988 Education Reform Act provided the legal background necessary for the first National Curriculum for Science, published in December 1988 – science for all had become a reality.

In practice, all children were now expected to study science for the whole of their compulsory schooling, building up to 20 per cent of curriculum time for most students aged 14–16. The examination system was revised to GCSE 'Double Award' science, in which three science subjects were condensed into the time required for two. Tiered papers for children of different intellectual needs provided differentiation for ability groups. A 'cut-down' single science version of the curriculum, a reduced version of the specifications for double science, was made available (primarily for those needing a less academic course, or wishing to study several languages to examination level). Those wishing to take separate examinations in the three separate sciences still took papers on the 'double science' specification, but supplemented by additional papers on the 'extra' biology, chemistry and physics. The era of 'one course fits all' had arrived.

Criticism of 'science for all'

'Science for all' can be regarded generally as a very positive achievement. All children experience science education throughout the years of compulsory schooling (5–16). Science is regarded as a 'core' subject in the UK along with mathematics and English. The entitlement of the population to receive science education is no longer disputed. However, 'science for all' has raised important questions. Paul Black notes that school science 'might be the only experience of serious learning about science that pupils have in their lives' (1993: 11). He raises the importance of factors such as appropriateness, accessibility, applications and authenticity, summarised as follows:

- children must understand and be confident about the science they study – this generates enjoyment, demands relevance and accessible, appropriate courses;
- the understanding must enable children to cope with issues arising in their own lives and in society, hence the science must include applications;
- the science must be authentic in terms of giving experiences of scientific enquiry which are realistic and do not assume scientific concepts developed smoothly;
- school science should contribute to children's intellectual and personal development;
- school science should provide a basis for making positive choices to continue with the subject further.

In some respects, 'science for all' as school science can be regarded as a victim of its own success. Although the main target of general accessibility and acceptance has been achieved, this has generated widespread and perhaps inevitable discussion about what the subject should now be like. The 'one course fits all', content-led curriculum based on old traditions linked with increasingly teacher-directed 'investigation' skills meets none of the criteria listed above and is regarded as unsatisfactory for most.

School science for specialists

The school science curriculum developed originally to contribute to an education for 'gentlemen', such that they were able to understand scientific developments, the process of SCIENCE and its contribution to society. As the demand for content grew and wider sections of the population gained access to education, school science became perceived increasingly as a source of providing educated young people destined for scientific careers. The grammar schools held great power in this regard, as only students who had attended these had the necessary qualifications to study science at university. SCIENCE continued to contribute significantly to the UK economy throughout the twentieth century with the rise of many science-based industries that are now household names, as well as academic departments with international research reputations. The demand for scientists has therefore remained strong. Two factors, the school system and industry's need for scientists, combined to create science curricula dominated by specialist provision.

This is how the situation has rested – all children have been subject to a mainly academic curriculum, the content of which has been determined by university entrance requirements or a group of mainly academics working at the behest of government. In many ways, the provision of science in schools has, apart from the first 50 years, been dominated by preparation of the next generation of scientists. Other concerns have played secondary roles.

The generalist v. specialist tension

Science is accepted as part of the core entitlement to education for 5–16-year-olds. In this respect, the 'generalist' provision has prevalence over the 'specialist' side. A look at the National Curriculum changes this perception, as the content can be seen as heavily based on old-fashioned academic science arising from grammar school curricula. Even with 'science for all', 'science for specialists' is still the reality. This has created significant problems. For example, a recent study by Jonathan Osborne and Sue Collins (2000) revealed that although 14–16-year-olds regard science knowledge as important in everyday life, many think of school science as hard, irrelevant, out-of-date, over-loaded with content, repetitive and boring to learn. Concern is expressed that too few girls, in particular, study physics, suggesting that gender bias is still prevalent. Osborne *et al.* (1996) indicated that less than 10 per cent of 16-year-olds choose science for further study.

On the other side, many schools have protested that the standard of scientific knowledge has declined significantly as 'science for all' has taken hold, such that academically and scientifically inclined children are no longer catered for. In particular, the development of GCSE 'Double Award Science' has received extensive criticism for failing to prepare students adequately for A level in terms of depth of knowledge and the level of understanding. The reform of 16+ examinations led to changes at A level and, not surprisingly, to arguments from the academic lobby that 'standards' have fallen here too. Universities responded by adding an extra year to science degree courses to create awards such as 'M.Phys.' and 'M.Chem.' in order, as it was put, to make up for the perceived knowledge deficit in new undergraduates. The implication is that school science provision for students wishing to specialise in science is now much weaker than previously and that this has had detrimental effects on the numbers and quality of students training to be scientists. The decline in recruitment to university science courses has led to the closure of some specialist physics and chemistry courses, while others have been combined with other subjects or changed titles to increase their appeal. The overall effect is regarded as putting the UK's SCIENCE research and industry base at risk, reducing national competitiveness in these areas.

The evidence indicates that the generalist v. specialist tension is not resolvable under the present system. Arguments will continue that academic students receive poor treatment, while non-academic children struggle with the frustrations of handling an out-dated, abstract curriculum. No one benefits from the curriculum being perceived as out-of-date and irrelevant. There is, therefore, a genuine tension between providing appropriate science education for the majority and the need to educate future scientists.

THE CHALLENGE OF CHANGE: HOW COULD SCHOOL SCIENCE DEVELOP?

There is wide agreement that the present school science provision does not satisfy the needs of the population. How school science can meet the needs of specialist and generalist, of those preferring content to process and vice versa, is debated. This section explores recent steps taken to address the issues raised and discusses areas that still remain unconsidered.

'Beyond 2000': a model for curriculum reform

During the late 1990s widespread frustration with the science National Curriculum and its assessment led several key science educators to propose a closed seminar series to discuss how science education could look in the twenty-first century. As part of the process, the participants returned to basic questions such as 'Why does an education in science matter?' and 'Who is school science education for?'. Consider your own answers to these questions before reading on.

The seminars generated a report, commonly called 'Beyond 2000' (Millar and Osborne, 1998), which is regarded widely as a blueprint for developing school science in the twenty-first century. The authors suggest that an education in science matters because it deals with major themes of importance and relevance to people, such as life and living things, how matter is organised and behaves, the Universe and so on. Education in science means gaining a foothold of knowledge about these things – specialist teaching is needed, because accurate information and understanding is difficult to gain through experience alone. Practically too, the authors suggest, science knowledge can help us make decisions about diet, health and lifestyle, contribute to scientific debate and understand science in the media. In answer to the second question, 'Beyond 2000' argues that science education is for everyone, and should be provided throughout compulsory schooling. Primary aged children are inquisitive and curious about their environments, and science can help to satisfy and stimulate this. The authors claim that the impact of SCIENCE is so great that the subject should be compulsory beyond the age of 14, so that everyone has met a minimum entitlement of information. School science education, however, must be regarded for the majority as an 'end-in-itself', providing a solid basis for future, preferably lifelong, learning and life as a citizen in a democratic nation.

'Beyond 2000' also presents an analysis of the problems inherent in the current provision. These, and the report's ten recommendations, are discussed next.

The problems of current school science education

Overtones of many of the problems raised in the report can be seen in our earlier discussion. The problems are summarised from the text of the report as follows:

- students appear to succeed in science, but their knowledge about 'outside world' science is very limited;
- school science does not develop and sustain children's intuitive wonder and curiosity about the world, expressed by older teenagers as failure to choose science post-16;
- the curriculum over-emphasises content, without showing the intellectual achievement behind many scientific ideas and the role of scientific evidence in constructing knowledge;
- the curriculum lacks clear aims and a model for how children's science knowledge may develop between the ages of 5 and 16;
- assessment is based on memory and recall, rather than being based on how science knowledge may be used in real life, such as analysing media reports;
- SCIENCE and technology are separated, although students perceive these as connected;
- contemporary science issues are not discussed;
- science lessons are often dull and uninspiring, as too little variety in teaching and learning experiences is provided and even investigations are becoming routine;
- the lack of choice post-14 does not take account of students' diverse needs, interests and talents (Millar and Osborne, 1998: 4–7).

Recommendations for change

'Beyond 2000' makes ten recommendations for a school science curriculum. These feature new ways of approaching school science teaching. We present the recommendations in full in Box 3.1. The report introduces three aspects new to a science education curriculum; scientific literacy, explanatory stories and ideas-about-science. These require some explanation.

By *scientific literacy* the authors mean achieving a population who are 'comfortable, competent and confident with scientific and technical matters and artefacts' (p. 9). Such a population would be able to follow new scientific stories presented in the media in a knowledgeable way, with understanding and with interest. Being scientifically literate would mean being able to express an opinion about social and ethical issues, as well as providing a basis from which retraining in science and technology could occur if this was required.

By *explanatory stories* the authors mean major ideas in science that contribute to our understanding of the world. Examples include the particle theory of matter, the gene model of inheritance and the helio-centric model of the Solar System. They contest that although the content of these stories is present in the current National Curriculum, the presen-tation prevents children and teachers from 'seeing' the big picture. Rather, the authors imply, the content gives the impression of being a 'pile of

---

BOX 3.1 RECOMMENDATIONS FROM 'BEYOND 2000'

1   The science curriculum from 5 to 16 should be seen primarily as a course to enhance general 'scientific literacy'.

2   At Key Stage 4 the structure of the science curriculum needs to differentiate more explicitly between elements designed to enhance 'scientific literacy', and elements designed for the early stages of a specialist training in science, so that the requirements for the latter do not come to distort the former.

3   The science curriculum needs to contain a clear statement of its *aims* – making it clear *why* we consider it valuable for all our young people to study science and *what* they should gain from the experience. These aims need to be clear and easily understood by teachers, pupils and parents. They also need to be realistic and achievable.

4   The curriculum needs to be presented clearly and simply and its content needs to be seen to follow from the statement of aims. Scientific knowledge can best be presented as a number of key 'explanatory stories'. In addition, the curriculum should introduce young people to a number of important ideas-about-science.

5   Work should be undertaken to explore how aspects of technology and the applications of science currently omitted could be incorporated within a science curriculum designed to enhance 'scientific literacy'.

6   The science curriculum should provide young people with an understanding of some key 'ideas-about-science', that is, ideas about the ways in which reliable knowledge of the natural world has been, and is being, obtained.

7   The science curriculum should encourage the use of a wide variety of teaching methods and approaches. There should be variation in the pace at which new ideas are introduced. In particular, case-studies of historical and current issues should be used to consolidate understanding of the 'explanatory stories' and of key ideas-about-science and to make it easier for teachers to match work to the needs and interests of learners.

8   The assessment approaches used to report on pupils' performance should encourage teachers to focus on pupils' ability to understand and interpret scientific information and to discuss controversial issues, as well as on their knowledge and understanding of scientific ideas.

9   In the short term, the aims of the existing science National Curriculum should be clearly stated with an indication how the proposed content is seen as appropriate for achieving those aims. Those aspects of the general requirements which deal with the nature of science and with systematic inquiry in science should be incorporated into the first Attainment Target 'Experimental and Investigative Science' to give more stress to the teaching of ideas-about-science; and new forms of assessment need to be developed to reflect such an emphasis.

10 In the medium to long term, a formal procedure should be established whereby innovative approaches in science education are trialled on a restricted scale in a representative range of schools for a fixed period. Such innovations are then evaluated and the outcomes used to inform subsequent changes at national level. No significant changes should be made to the National Curriculum or its assessment unless they have been previously piloted in this way.

(Millar and Osborne, 1998)

bricks', rather than a significant building. Their thinking is that by organising the curriculum into a series of narratives aiming to communicate key ideas, children will remain engaged and interested, as the content will appear more 'coherent, memorable and meaningful' (p. 13).

By *ideas-about-science* the authors mean that students should learn about the 'scientific approach to inquiry' (p. 19), that is, how scientific data are obtained and the problems and limitations associated with these data. They suggest students are presented with information exploring risk, probability and links so that reasoning about a situation can be rehearsed and understood. Students also need to understand the social processes involved in SCIENCE, which scrutinise knowledge claims before their acceptance and recognise how external factors can influence knowledge.

These suggestions are at the heart of the report's proposals, placing quite a different emphasis on the role and place of school science in society from the present situation. The two tensions described earlier become mainly irrelevant under these recommendations, because the dependence on content knowledge based on the needs of specialists is removed and the debate from content versus process is changed to a more general discussion about scientific literacy, stories of SCIENCE and scientific evidence. Further, the proposals offer a means of resolving the specialist versus generalist tension by offering different curricula during Key Stage 4, but within the 20 per cent time allowance provided currently. Everyone would take a standard compulsory science course occupying 10 per cent of curriculum time, while the remaining 10 per cent would permit students to choose courses from a wide range of options, such that a mix of academic and less academic needs could be satisfied. This choice of options, they propose, would continue throughout the 14–19 age range, so that students who realise at a later stage that they would like to pursue a scientific career would not be excluded by having made inappropriate option choices at Key Stage 4.

Giving the science curriculum clearly stated aims based on living in a modern society is a pragmatic suggestion which would bring the UK curriculum into line with those of other countries such as New Zealand (New Zealand Ministry of Education, 1993). The aims would help to maintain science as a high-status subject alongside numeracy and literacy. In this respect, the authors are making claims for science not dissimilar to those made by the British Association in 1917.

The recommendations also suggest making explicit connections between SCIENCE and technology. Osborne and Collins (2000) indicate that students often place SCIENCE and technology in the same 'frame', regarding technological progress as being dependent on scientific ideas. By making this recommendation, the authors stress the impact technology has on our lives, and particularly in relation to the applications of scientific ideas. For example, the technological advances which made satellite communication, the sequencing of the human genome and heat-sensitive materials possible are rarely discussed in school science.

UK schools are focused heavily on assessment so, unsurprisingly, the report draws attention to a need to alter significantly how students' learning and progress in science may be judged, if the curriculum were to change along the suggested lines. Incorporating up-to-date newspaper articles and other scientific reports, data analysis, the role of scientific evidence and discussion about scientific phenomena in assessment would require a major re-think of strategies. The outcomes of such a review may be very positive in terms of removing the dependence on rote learning and reducing time pressures.

What might a school science course based on 'Beyond 2000' look like?

Since the publication of 'Beyond 2000', a new GCSE course called 'Twenty First Century Science' (www.21stcenturyscience.org, accessed 1 August 2004) started trials in September 2003. This course, developed by the Nuffield Foundation and the University of York Science Education Group, offers a 'three in one' set of GCSE science choices, based on the 'Beyond 2000' model. One course, taking 10 per cent of curriculum time corresponds to one GCSE and is taken by everyone. During the remaining 10 per cent of science time, students can choose a second GCSE called Additional Science (Applied) or Additional Science (General). The core science course comprises nine modules based on themes of interest to teenagers, such as air quality, material choices, you and your genes and the Earth in the Universe. The 'Applied' and 'General' courses comprise five 50-hour modules of which students choose three. The Applied modules take a vocational outlook, emphasising work contexts such as management, communicating knowledge and application of knowledge to solve problems and analyse data, for example. The General course comprises nine modules with no choices. The outlook here is on preparing students for post-16 science study.

Vocational GCSEs (DfES, 2003c) have also been made available since September 2002 in a wide range of subjects including science. The vocational science GCSE (VGCSE) has aspects that fit with the 'Beyond 2000' vision for science education, in that the course aims to offer a more 'hands-on' approach to learning, emphasising practical skills, application of knowledge and skills and the use of knowledge in the workplace. The aim of these courses is to stimulate an interest in learning among students who may struggle with a traditional, academic approach, by introducing them to science industries, developing understanding of key concepts and developing their skills. The VGCSE Science offers students the opportunity to develop specific key skills such as problem solving and working with others and to experience a wide range of teaching strategies such as joint and individual project work, visits to industry and use of case-studies. Assessment includes a portfolio that students collect and develop during the two years.

These two developments are at early stages and no formal evaluations are available at present to help analyse their success in achieving their aims. However, clearly both attempt to offer something very different from the standard Double Award GCSE package that has received such criticism. We discuss both courses in more detail in Chapter 4. Having looked at the most recent new developments, let us now consider what more remains to be done.

A critique of 'Beyond 2000': what should school science include?

'Beyond 2000' has enjoyed significant impact as a way forward for science education for the twenty-first century. The new GCSE course, 'Twenty First Century Science' is putting the 'Beyond 2000' recommendations into practice, while the DfES itself has also acknowledged a need to extend the range of GCSE provision beyond the usual curriculum. However, closer inspection is warranted, because there are areas and aspects of science that are significant by their absence from both the recommendations and the new courses.

First, 'Beyond 2000' makes no attempt to address creativity in science education, focusing instead on an entirely rational-based view of SCIENCE. Science remains outside the general development of creativity in education. School science is still perceived as a way of indoctrinating students to the rational view of the Universe, embracing traditional science philosophy, social constructivism and reliable knowledge. In contrast, the recent report 'All our futures' (NACCCE, 1999) emphasises creativity in learning, defining it as: 'An imaginative activity fashioned so as to produce outcomes that are both original and of value' (p. 28).

Imagination, speculation and initiative in science are given no real place in a curriculum proposal based entirely on a rational view of the subject. This could mean that students with creative talents may be excluded from good performance, but perhaps more seriously, that science begins to lose ground against other subjects that have adopted (or will adopt) a more open-minded view towards the role of creative thinking skills in the learning process.

Second, the view of SCIENCE presented in 'Beyond 2000' is still largely that relevant to a Western culture in which scientific developments regarded as 'major' have mainly been made by men working in laboratory (or similar) settings. Reiss (2002a), in an article unrelated to 'Beyond 2000', makes this point very effectively by describing a set of postage stamps published by the Royal Mail in 1991, depicting 'Scientific achievements'. The four chosen were Michael Faraday (electricity), Charles Babbage (computer), Robert Watson-Watt (radar) and James Whittle (jet engine). Reiss comments:

> I find it difficult to imagine a narrower conception of what science is and who does it. The image seems to be that real science is hard physics, with military applications, done by males who are white and worked on their own between 1820 and 1940. No wonder so many students drop science at school as soon as they have the chance!
>
> (Reiss, 2002a: 40)

'Beyond 2000' does not specifically state that SCIENCE developed outside the Western world *should* form part of the UK school curriculum, nor that the achievements of women scientists should be promoted alongside those of men, nor that the conception of what SCIENCE is should be extended beyond the traditional laboratory-based view. The 'explanatory stories' focus on scientific achievements made by males and indicate no breadth of cultural background. In a second article, also unrelated to 'Beyond 2000', Reiss (2002b) suggests that offering a curriculum based on narrow, male-dominated, Western-oriented cultural values is not only alienating for many children, but is also offering them an 'impoverished' reflection of SCIENCE as school science. Reiss cites examples such as surgery, magnetism, personal health, and iron and steel as topics to which non-Western cultures have contributed significantly. There is a need, we contest, to offer school science that is enriching for our students, rather than just a re-packaging of the traditional aspects of the subject that have formed the basis for school science since its inception.

A third, related criticism is that 'Beyond 2000' still does not really offer a clear statement about what constitutes SCIENCE. To help clarify our meaning, we can turn again to Reiss (2002a), who illustrates this point with his account of women, some untrained, who studied animal behaviour rigorously and insightfully over long periods of time. Their observations led to significant developments in understanding of this scientific discipline, yet other scientists disregard their work as not being 'true science', for various reasons. The point is that we need to be wary of presenting a picture of what SCIENCE 'really is' – in Reiss's view, the methods of SCIENCE as presented in school science are much more narrow than they need or should be.

WHAT DO YOU THINK?

What we seek, perhaps, is the most comfortable option – school science that sits most easily in our Western culture and, preferably, is closest to the 'traditional' SCIENCE with which teachers, as trained scientists, are familiar.

However, we need to go beyond this to look at the children we are teaching and the world they are likely to inherit – already shrunk by near instant global communication of various picture and text forms; in which distance can be travelled easily and long life and good health a realistic expectancy for many in our society. Is it therefore correct to teach a curriculum to suit us, the present generation of those wielding the power and influence? Or should we take the courage to step outside our own comfort zones in the anticipation of providing something more than we ourselves experienced in school science for the possible future benefit of those we teach?

THINKING ABOUT PRACTICE

■ Why does an education in science matter?

■ Who is school science education for?

■ What practices, in your view, should reasonably be called 'SCIENCE'?

■ Using your answer to the previous question, what aspects of SCIENCE do you think are vital for school students to experience? Justify inclusion of each in your own science curriculum proposal.

Chapter 4

New perspectives on science education

Without experiment I am nothing; still try, for who knows what is possible?

(Michael Faraday)

INTRODUCTION

In this chapter, we describe and discuss perspectives on science education that have arisen in the last ten years or so. Inductees may experience, or at least hear about, one or more either during a teaching placement or through an ITT course, so besides serving the function of providing information, we also include this material to help prompt further reflection on how school science could be shaped in the future.

In Chapter 2, we explored the notions of school science in relation to SCIENCE and the distinct science disciplines. We conclude that pinning down any essence of SCIENCE in terms of subject matter (science can be 'about' just about anything), or even in terms of a singular scientific method, is problematic. We explored these issues at some length because we think inductees need to understand in what ways 'school science is not SCIENCE'. This issue is, to us, fundamental to the process of becoming a professional science teacher. Our conclusion was that the common ground among the sciences, and potentially between SCIENCE and school science, can best be found in terms of a 'scientific attitude' that can be reflected in terms of a set of aims, such as those suggested by the ASE in its policy on Values and Science Education (Box 2.2, p. 37).

In Chapter 3, we pointed out that although the science curriculum may be mandatory in state schools, the curriculum is socially constructed through a political process, and so can be reconstructed in the future.

A worthwhile consideration is how we would wish to reshape the curriculum. In fact, at the time of writing there are a number of exciting developments taking place. In this chapter we discuss some recent and ongoing developments at KS3, KS4 and post-16.

NEW PERSPECTIVES AT KS3: THE NATIONAL STRATEGY FOR SCIENCE

In 2001 the UK government began a National Strategy to improve the quality of children's learning at KS3. One reason for this is that concern was being expressed that children's results in the KS3 SATs were not improving as anticipated. English and mathematics featured in the initial stage of the Strategy, with science joining in 2002. The Strategy was designed to help schools implement and teach the NC. The NC documents in all subjects can be found at www.nc.uk.net (accessed 1 August 2004). Here we introduce the main points associated with the KS3 National Strategy for Science.

Outline and purposes of the KS3 National Strategy: Science

KS3 comprises the first part of secondary education and is designed for children aged 11–14. A number of educators have expressed the view that this age group is very significant in terms of promoting interest in science – if children can be captivated, enthused, challenged and stimulated during this critical stage, it is thought that they are more likely to develop and maintain a positive attitude towards science during their mid- and late teens.

The KS3 Strategy forms part of the government's curriculum 'guidance', designed to improve standards. Science, as a core subject in the NC (see Box 5.1, p. 96), has its own strand of the Strategy. The Strategy provides a 'Framework' for teaching science at KS3 for schools to use which is available from the DfES standards website at www.standards. dfes.gov.uk/keystage3/respub/scienceframework/ (accessed 1 August 2004). A Continuing Professional Development (CPD) programme provided by a system of consultants working in LEAs supports the implementation of the documentation, with the expectation that schools send representatives to training who will 'cascade' the information and ideas back to colleagues within their science departments. The Framework is designated for guidance only, so schools are free to ignore this officially-recommended-but-not-mandatory package. The purposes of the Framework are given in Box 4.1.

The Framework sets out teaching objectives for the three KS3 years, based around a notion of 'key ideas'.

Key ideas identified in the KS3 Science Framework

One interpretation of the Framework is to help teachers transform the curriculum more successfully into lessons that provide better coherence and continuity for students. The main way of doing this is to identify a set of 'key ideas'. This seems to us a sensible approach, as a limited number of key ideas can act as organising themes for teachers when conceptualising their teaching, and as identifiable threads for learners trying to make sense of the learning. There is a great deal of science to be learnt in the first three years of secondary education, and it can be

---

BOX 4.1 PURPOSES OF THE FRAMEWORK FOR TEACHING SCIENCE

The KS3 Science Framework aims to:

- bring together in one place the experience of the pilot and best practice in secondary schools . . . ensure that scientific enquiry is integrated with and taught alongside knowledge and understanding in a range of contexts;
- identify the key scientific ideas that underpin science at Key Stage 3;
- set out yearly teaching objectives for years 7, 8 and 9 that build on science in Key Stage 2 and develop pupils' understanding of the key scientific ideas in Key Stage 3;
- give advice on how teachers and trainee teachers can . . . plan and teach appropriately challenging and engaging work to their pupils;
- provide a basis for target setting;
- enable headteachers and curriculum managers to set high and consistent expectations for pupils' achievement.

(DfES, 2002a: 8)

---

argued that there is too much content to provide students with a coherent experience of science. By using the key ideas as 'touch stones' within the curriculum, the Framework was designed to show how there may be progression and continuity (see pp. 66–8) as students develop increasing understanding and mastery of the key ideas.

The teaching objectives set out in the framework are clustered around the five concepts which (along with scientific enquiry) are identified as key ideas:

- energy;
- particles;
- forces;
- interdependence;
- cells.

The Framework is designed around showing how there can be continuity and progression as students develop increasing understanding and mastery of the key ideas. Each of these five 'key ideas' is certainly important in helping children gain understanding of aspects of SCIENCE, and all deserve considerable emphasis in school science. It is open to debate, however, whether these are the most useful foci for conceptualising secondary science teaching. We will take each in turn and consider their suitability as candidates for 'key ideas'.

Energy and *particles* can certainly be used as explanatory ideas throughout secondary science and beyond and we have no qualms with them being identified at this level.

Energy is a consideration in every school science topic, yet is difficult to teach effectively due to its abstract nature. Finding an authentic way of teaching about energy that makes sense to 11-year-olds is a challenge that the physics education community has struggled with (e.g. Driver and Millar, 1986; Solomon, 1992; Warren, 1983). However, the Strategy does both recognise the central role of energy as a science concept and provide teachers with guidance about teaching it, which is welcomed.

Particles is also an essential concept, as all scientific disciplines draw upon the molecular model of matter. Although many KS3 topics can be taught without applying this model, much school science can only be explained satisfactorily in the context of the particle model. Good explanations are central to SCIENCE, so we think it necessary that science teachers use the most powerful explanatory devices available, transformed to make sense to learners. Again, this is a serious pedagogic challenge for which we welcome any good guidance teachers can be given.

We are less convinced, but willing to be convinced, that *forces* can act as a key idea underpinning KS3 science. We would, however, expect progression in understanding across KS2, through secondary education and beyond. Research shows that by years 12–13 students may have compartmentalised ideas relating to fields and forces as 'physics' knowledge, instead seeking non-physical explanations for chemical processes (Taber, 1998a, 2003c). This suggests that 'force' is a concept area that needs more emphasis in teaching some aspects of chemistry in the secondary curriculum. However, we are not sure that *forces* can claim to be as useful a key idea across SCIENCE as energy or particles.

The concept of *interdependence* has potential as a key idea, being relevant in topics such as gas laws, rates of flow, ecology and chemical equilibrium. Much SCIENCE is about modelling aspects of the world complicated by different types of *interdependence*. Within the Framework, *interdependence* is seen in the more limited context of 'interdependence in and between biological and physical environments', which seems a much more narrow definition than would be most useful.

We are even less convinced about the status of *cells* as a key idea across KS3 science. This concept is important in Sc2 (life processes and living things – see Table 2.1, pp. 22–3), but this idea will not provide structure and organisation to student learning across the whole of science.

So, while we consider the principle of identifying 'key' ideas as sound, we are only fully convinced about two of the five candidate key ideas suggested in the Strategy. We feel that it would have been possible to identify a set of key ideas that are better able to permeate and support learning across science.

Progression and continuity: making connections with KS2

The *key ideas* should take a *key role* in ensuring continuity and progression in student learning. Progression and continuity are two significant terms important for inductees to get to know. Definitions may vary slightly, but in this context, the Framework document defines continuity as: 'consistency in expectations and teaching approaches between and within key stages. Good continuity extends pupils' experiences without unhelpful repetition.'

And progression is defined as: 'the step-wise development of scientific concepts and techniques, though the steps are not necessarily equal in size or taken at regular intervals'.

Progression is a major concern of the Strategy, with a specific guidance pack focusing on 'Planning and implementing progression for science in the classroom' (DfES, 2002b). For example, Box 4.2 reproduces advice on interpreting the NC levels through which it is expected students will pass as they make progress in their science learning.

The Strategy is particularly concerned to ensure that children make progress in science from one Key Stage to another. Making a smooth transition between Key Stages is very important, especially when we recall that the majority of children transfer between schools between KS2 and KS3. Science teachers achieve this by ensuring good sequencing of tasks and providing good support to learners when introducing new ideas. We explore the 'how to' of science teaching in more detail later in the book.

When progression is recognised as being of such importance, it is a concern that, as the Framework concedes, 'areas of Year 7 science that are not closely related to the Key Stage 2 programme of study' (DfES, 2002a: 23) include:

- cells;
- particles;
- chemical reactions;
- energy.

BOX 4.2 GUIDANCE ON JUDGING PROGRESSION WITH THE NC LEVELS

'Main characteristics of levelness' (what students are expected to show/do) as suggested in the KS Strategy

Level 3   Simple explanation of what
Level 4   Correct terminology
Level 5   Applying concrete ideas to less familiar situations
Level 6   Using abstract ideas to explain why things happen
Level 7   Linking ideas from different areas; using quantitative relationships

(From DfES, 2002b: 55)

So there is little preparatory work in the primary school on which to base teaching about three of the five key ideas identified at KS3.

It is interesting that 'chemical reactions' features on this list. We feel that *chemical reactions* has at least as much claim to be a key idea in science as *cells*. We wonder if the Strategy could have been a little more adventurous in identifying its key ideas, and looked for ideas which are more truly fundamental to the science curriculum, such as:

- building blocks (which could include atoms, molecules, cells etc.);
- change (including cyclical changes);
- conservation (including, but not limited to, energy); and
- interdependence (in its widest sense).

Perhaps these ideas may look too abstract for students at Key Stage 3, but this does not matter if their purpose is to inform the teacher in the process of transforming the curriculum through scheme of work and lesson plan to classroom action.

The KS3 Strategy model

Box 4.3 summarises the expectations made of participating schools by the architects of the KS3 Strategy. There is an expectation that schools will nominate staff to be involved in CPD opportunities on behalf of departments or the school. In some ways, this is an exciting model. Its success depends on:

- consultants having the energy and vision to create exciting training rather than just 'deliver' the units;
- teachers attending the training and having time to work with the materials to cascade the ideas back to the department; and
- time being available in science departments to allow feedback to take place.

The model is based on the arrangement that a science teacher would participate in the training – for example, to strengthen the teaching of cells and then spend time reporting back to the rest of the science department. This is potentially sensible. We await some form of independent evaluation to see how well the model has actually been implemented, but from what the authors hear when visiting schools, the implementation

BOX 4.3 KS3 STRATEGY – EXPECTATIONS OF SCHOOLS

As part of the Strategy schools are asked to:

- audit standards, teaching and learning;
- make effective use of the Strategy's Frameworks for teaching and the QCA schemes of work, either directly or to customise their own schemes of work;
- take part in selected units of continuing professional development and undertake follow-up work in school;
- take part in whole-school initiatives on cross-curricular issues, such as literacy and numeracy across the curriculum;
- support transition from Key Stage 2: for example, by offering a summer school, and by providing catch-up classes for Year 7 pupils who did not previously achieve level 4 in English and mathematics;
- provide mentoring for Year 8 pupils who are falling behind and becoming disaffected;
- provide booster support for Year 9 pupils before the national tests for Key Stage 3.

(DfES, 2002: 7)

has, in practice, been patchy. However, given that the Framework has the status of 'guidance', so participation is not compulsory, this is perhaps to be expected.

The DfES/QCA Scheme of Work for KS3 Science

The Framework suggests that schools also consult another document produced by the Government in conjunction with the QCA. This is called the Scheme of Work (SoW) for KS3 (available online at www.standards.dfes.gov.uk/schemes2/secondary_science/, accessed 1 August 2004).

This document is commonly referred to as the 'QCA Scheme of Work' and has been a little controversial. The document states that:

This scheme shows how the science programme of study for key stage 3 can be translated into manageable units of work.

The scheme is not statutory; you can use as much or as little as you wish. You could use the whole scheme or individual units. The sections on 'using this scheme' and 'sequencing the units' can help you to decide how you want to use the scheme.

(From www.standards.dfes.gov.uk/schemes2/
secondary_science/, accessed 1 August 2004)

Table 4.1 gives the units in the QCA SoW. As the information suggests, the aim of the SoW is to provide help and advice about how to make the NC Science document work in practice. Clearly, the guidance is not intended to be statutory – schools cannot be legally bound to teach in the way the QCA SoW suggests. However, when coupled with a financially beneficial CPD programme related to a National Strategy Framework, the document becomes rather more powerful. The Government appears to be introducing defined ways of teaching which will help to ensure that children in schools meet the necessary standards. By providing such a document, school science departments and their professional teachers, are almost being denied the opportunity to create their own schemes. Although we agree, of course, that the purposes of the Framework, and indeed the QCA SoW, are certainly worthy, we wonder if the QCA SoW and the National Strategy Framework will remain non-mandatory, or eventually become 'statutory' alongside the NC. Science teachers should be professionals and not educational technicians. We consider that one strong feature of becoming a professional science teacher is to be able to exercise one's own judgement and creativity in planning work suitable for the learners presented in each teacher's classroom. This, to us, is a vital part and perhaps one of the most enjoyable aspects of the job.

Teaching and learning strategies

Before leaving the topic of the National Strategy for Science at KS3 it is worth briefly exploring the way in which teaching science is approached in Strategy materials. The Framework recommends what is described as a 'focus on direct, interactive teaching' (p. 41), using activities (pp. 42–3) such as:

- directing and telling;
- demonstrating;

Table 4.1 *Units in the QCA Scheme at KS3*

Year 7 units	Year 8 units	Year 9 units
7A Cells	8A Food and digestion	9A Inheritance and selection
7B Reproduction	8B Respiration	9B Fit and healthy
7C Environment and feeding relationships	8C Microbes and disease	9C Plants and photosynthesis
7D Variation and classification	8D Ecological relationships	9D Plants for food
7E Acids and alkalis	8E Atoms and elements	9E Reactions of metals and metal compounds
7F Simple chemical reactions	8F Compounds and mixtures	9F Patterns of reactivity
7G Particle model of solids, liquids and gases	8G Rocks and weathering	9G Environmental chemistry
7H Solutions	8H The rock cycle	9H Using chemistry
7I Energy resources	8I Heating and cooling	9I Energy and electricity
7J Electrical circuits	8J Magnets and electromagnets	9J Gravity and space
7K Forces and their effects	8K Light	9K Speeding up
7L The solar system and beyond	8L Sound and hearing	9L Pressure and moments
		9M Investigating scientific questions

- explaining and illustrating;
- questioning and discussing;
- exploring and investigating;
- consolidating and embedding;
- reflecting and evaluating;
- summarising and reminding.

We would support the range of strategies listed, advocating that variety and quality are also essential parameters – a lesson which comprises 'directing and telling' for 60 minutes is, unless in very special circumstances, unlikely to challenge children appropriately.

The Framework also includes a sample lesson plan (pp. 38–9). This states the lesson objective, vocabulary and resources, and divides the lesson into three main sections:

- starter activity (10 minutes);
- main activity (40 minutes – further divided);
- plenary (10 minutes).

According to the Framework, lessons should 'begin with setting the scene and a short activity to help pupils tune in, interest them and engage their attention' (p. 44). The main activity should build upon the starter activity. The concluding plenary should be 'dynamic', and should 'help the pupils to reflect on the lesson, say what was important about it and consider the progress they have made' as well as 'anticipate what the next steps might be' (p. 45). This three-part lesson structure is a feature of the KS3 Strategy (not just in science), although the framework claims that this 'is not a mechanistic recipe' and that 'professional judgement' should be used to determine actual timings and activities.

Comments and critique on the KS3 National Strategy: Science

Undoubtedly, the National Strategy sets out to support teachers, providing much worth considering and reflecting on (see Chapter 10). We encourage teachers to access the Strategy materials and engage with them critically – in other words, look closely at the ideas offered with an open mind to see which may be adopted or modified as part of a teaching repertoire.

We think, however, that the overall flavour of the Strategy is in some ways a mix of trivial (for example, qualified teachers should not really need to be told they can 'tell', 'direct' and 'question'); the inadequately considered (such as the choice of key ideas); and the 'genuinely-useful-but-likely-to-be-lost' – those 'gems' likely to be missed due to the sheer quantity of the materials. To conclude this brief overview of a substantial initiative, we will just refer to three examples of the 'treasures', which are rather 'buried' in the Strategy materials.

The guidance on progression, *Planning and implementing progression for science in the classroom* (DfES, 2002b) refers to Bloom's taxonomy of educational objectives. However, rather than discuss the revised taxonomy which draws on the benefit of decades of application of the original ideas (see Chapter 6), the Strategy just uses the somewhat dated 1964 version.

Second, in the materials designed to 'strengthen' the teaching of energy at KS3 (DfES, 2003b) there are two papers among the appendices written by academics who are among the UK's most respected names in science education internationally. Robin Millar provides a paper outlining the conceptual issues on teaching energy ('Teaching about energy', pp. 101–19) – but this is assigned to an appendix, and does not seem to have influenced the training materials themselves.

Third, another appendix ('Types and forms of model', pp. 95–7), is written by John Gilbert, probably the world's foremost expert on models and modelling in science education. This is a short précis of some of the key ideas about the roles of models in thinking and teaching. In view of the importance of models in SCIENCE and in the curriculum we would expect this material to be drawn upon significantly in the Strategy. The Framework does refer to the use of models and analogies in teaching (p. 15), but this material only talks about students using and critiquing models which have been provided for them. We would very much like to have seen students being taught about the ways in which models are used in SCIENCE and encouraged to develop their own. This seems another missed opportunity.

DEVELOPMENTS AT KS4: NOVEL GCSE COURSES

Curriculum development has been a strength of the UK education system since the 1960s. Prior to the introduction of the NC, science teachers had great professional freedom to both choose the most appropriate examination course for their students and, if they wished, to be closely involved in the development of new courses. The introduction of the NC and subsequent reorganisation of the examination system has created less flexibility, with a reduction in the number of examination boards and the range of courses available. However, this has not dampened enthusiasm for developing new GCSE and A level courses, as well as materials to support teaching and learning. Here we describe novel approaches to GCSE that have arisen since 2000.

Twenty First Century Science – three GCSE courses aiming to educate the citizen

This approach to GCSE science has been commissioned by the QCA using funding from three charities – the Nuffield Foundation, the Salters Institute and the Wellcome Trust. The project, directed by John Holman, Robin Millar and Andrew Hunt, is developing three new courses for KS4 to provide variety and diversity such that students' different needs can be met. The key aim of the courses is to provide school science for the citizen, recognising that everyone requires a core entitlement to understand aspects of science, while some students will wish to specialise. At the time of writing, 'Twenty First Century Science' or 'C21 Science' as it is called, is being trialled in 80 schools in England and Wales. As we mentioned in Chapter 3, the project is directly related to the recommendations of the 'Beyond 2000' report (Millar and Osborne, 1998), featuring 'Science Explanations' and 'Ideas-about-Science'.

The three courses are called 'core science', 'applied science' and 'general science'. The idea is that everyone takes core science, which explores issues and topics from the perspective of a member of the public. Students can then choose if they wish to take an 'additional' science GCSE. In the C21 Science scheme, this may comprise 'applied' or 'general' modules. The titles of the modules are shown in Table 4.2.

The general modules are designed to prepare students for post-16 science study, while the applied modules take a more employment-related approach, adopting the perspective of SCIENCE practitioners. Students taking the general course must study all nine modules. Those taking the applied course must choose three from the six available.

The core science course aims to produce young people who are scientifically literate. The course developers define scientific literacy as being able to:

- appreciate and understand the impact of science and technology on everyday life;
- take informed personal decisions about things that involve science, such as health, diet, use of energy resources;
- read and understand the essential points of media reports about matters that involve science;
- reflect critically on the information included in, and (often more important) omitted from, such reports; and

Table 4.2 *Modules in 'Twenty First Century Science' GCSE courses*

Core Science GCSE modules	Additional Science GCSE	
	Applied modules	*General modules*
Air quality	Life care	Homeostasis
You and your genes	Products from organisms	Chemical patterns
The Earth in the Universe	Scientific detection	Why things move
Food matters	Harnessing chemicals	Growth and development
Radiation and life	Communications	Chemicals in the natural environment
Material choices	Materials and performance	Modelling the behaviour of electric circuits
Keeping healthy		Brain and mind
Radioactive materials		Synthesis and analysis
Life on Earth		The wave model of radiation

- take part confidently in discussions with others about issues involving science.

> (Quoted from website, www.21stcenturyscience.org/ newmodel/literacy.asp, accessed July 2004)

This was a key theme of the 'Beyond 2000' report, placing emphasis on skills that have not featured strongly in school science previously. Developing scientific literacy requires that students are trained to think in a different way from simply learning facts about a topic. In this course, the notion of 'Ideas-about-Science' is used to organise the way scientific literacy is taught. Using this notion, the developers wish to help students reflect on scientific knowledge, including:

- the practices that have produced it;
- the kinds of reasoning that are used in developing a scientific argument; and

- the issues that arise when scientific knowledge is put to practical use.

(Quoted from website, www.21stcenturyscience.org/
newmodel/literacy.asp, accessed 1 August 2004)

They place specific Ideas-about-Science into six categories:

- data and its limitations;
- correlation and cause;
- theories;
- the scientific community;
- risk;
- making decisions about science and technology.

(Quoted from website, www.21stcenturyscience.org/
newmodel/literacy.asp, accessed 1 August 2004)

The second key recommendation from 'Beyond 2000' was that scientific explanations should be used to portray 'big ideas' in science, rather than, as at present, teaching small pieces of detail presented in formats prohibiting students from understanding or 'seeing' an overall picture. Accordingly, a selection of big ideas is used to provide a framework for 'making sense of the world'. One point of the C21 Science course is to select a few explanations viewed as of central importance to this, without becoming absorbed in overwhelming detail. The ideas selected by the course team are:

- chemicals;
- chemical change;
- materials and their properties;
- the interdependence of living things;
- the chemical cycles of life;
- cells as the basic units of living things;
- maintenance of life;
- the gene theory of inheritance;
- the theory of evolution by natural selection;
- the germ theory of disease;
- responding to stimuli;
- energy sources and use;
- radiation;

- radioactivity;
- the Earth;
- the Solar System;
- the Universe.

(Quoted from website, www.21stcenturyscience.org/
newmodel/literacy.asp, accessed 1 August 2004)

Comments and criticism about C21 Science

We can see much that is good in this curriculum model. It is especially admirable to see ideas in a report translated into practice. The notion of educating students to be scientifically literate in the sense meant by the course developers is powerful, as well as developing understanding in key science areas. To be successful, this will require carefully designed course materials and trained teachers as well as a sensible, valid means of assessment in order to realise the outcomes in a meaningful way.

The scientific explanations list bears some criticism. While we cannot deny the significance of the listed topics, the balance between three mainstream science subjects appears a little awry – we identify four that may be regarded from the titles as 'chemical', seven 'biological' and four 'physical' and, without further explanation, it is difficult to think simply of 'chemicals' as a major idea alone. We assume that these explanations will be taught throughout the modules listed in Table 4.2. It is also difficult to perceive much information from the titles alone, but it could be that when C21 Science goes 'live' into schools, delivering the material may be as problematic as the current situation. By this we mean that science teachers may still be required to work as experts in areas outside their normal competence, offering a professional development opportunity for developing new skills, but also causing a reduction in the quality of science education being offered. Teachers would also definitely require different and new skills from those currently promoted in order to successfully deliver this course. For example, experience in the pilot has shown the coursework components of the course expect students to demonstrate 'research' skills that they will not have developed through their KS3 studies (Reed, 2004).

However, it may also be that the materials are more closely in line with teachers' own experiences of being scientists, so former professional identities may be more directly useful when teaching C21 Science than they are using a traditional GCSE Science course.

The choice element may also prove to be problematic. For example, some secondary schools have a very strong bias towards offering academic courses, so may not offer all three courses to their students. While this in itself is perhaps not problematic, students changing schools may experience difficulties – a student who has followed an applied course may find that this is no longer available to him/her on moving to a strongly academic school. Students taking the applied course may find that module choice is restricted by staff availability and, on changing schools, that their original choices are not available. If C21 Science is to be the model for the future, access to all aspects of the course are needed for everyone.

Assessment may be a third issue to be resolved. C21 Science will attempt to assess scientific literacy skills, perhaps using novel approaches at this level such as 'open book' examinations and current media stories. This can be commended, although finding valid and reliable methods for this may be problematic. The course developers say, at the time of writing, that the assessment model is under preparation. However, there is also an issue with the core science and applied/general science courses. The core science course will be available to be taught in one year, so it is possible that schools may organise their teaching so that all year 10 students take this first and then are invited to choose what science, if any, to take in year 11. This would mean students taking a formal, terminal assessment in each of years 10–13, assuming their participation in post-16 courses. Alternatively, students may take two courses at the same time, so having lessons in core and applied or general science throughout the two years of KS4. Depending on the organisation of course material, this may prevent students from benefiting from the core units before tackling the applied and general modules. It remains to be seen how schools will implement this model.

C21 Science has much to commend it as a curriculum model for school science. However, we also draw readers' attention to the comments made in Section 2.6 highlighting aspects of SCIENCE that appear to be absent from 'Beyond 2000'. If this course represents a real improvement over what has gone before in the fullest possible sense then inclusion of at least some of the issues raised in Chapter 3 (pp. 58–60) is necessary. Otherwise, C21 Science could end up being a re-package of school science taught for many years, but simply with new flavours added.

Applied Science GCSE: part of the 14–19 reform

The government has sought to raise standards in education for 14–19-year-olds, as well as KS3 children. The DfES paper '14–19: Opportunity and Excellence' (DfES, 2003c) set out the government's case for reforming education in this age group, the aims being to promote higher participation and attainment; develop a system capable of including generalist and specialist studies; and, more generally, promote development of young people equipped for employment, life and lifelong learning. As part of this reform, new GCSE courses have been developed in eight subjects including science. Here we provide more information about this course.

GCSE Applied Science has been available since 2002 as a 'double award' course equivalent to two GCSEs. The first students are just completing the new courses. The 'vocational' GCSEs, as they are called, are designed to be available both in secondary schools and post-16 colleges. This is to encourage the widest possible participation. Once the course is made available in their school, 14-year-olds may select Applied Science as an alternative to the traditional GCSE Double Award Science course. Sixteen-year-olds may choose the course as part of their post-16 programme. GCSE Applied Science students are taught in a work-related way. Materials to support the course are available (for example, Topham *et al.*, 2002). These include a student's book, a teacher's guide and a revision and coursework book. The course provides an alternative to a traditional format GCSE in several different ways.

First, the course aims to include a wider range of learning opportunities, including visits to industry and/or business, project work, world wide web-based research, use of case studies and role play/other simulations, such as simulating a business/research opportunity. The objective of these is to find out more about how people work in SCIENCE and how science is used in work situations.

Second, assessment will include a 'portfolio', as well as externally set terminal examinations. This means that during the course students will assemble pieces of work reflecting their attainment and the range of activities undertaken. The precise requirements for the portfolio will be determined in the specification. The portfolios will be subjected to moderation to ensure consistency of standards between the schools and colleges delivering the course.

Third, the course is designed such that students are encouraged to progress to a higher level qualification such as a VCE (vocational

A level), or an AS course in a closely related subject. Other alternatives are also possible – students may go on to a diploma, NVQ, or modern apprenticeship.

Fourth, students will be encouraged to undertake work experience as part of the course. This means that students will spend some time 'at work' in a science-based business or industry, perhaps for example, assisting in an optician's, working in a pharmacy, working in a research laboratory, health clinic, hospital, the chemical industry or a food manufacturer. Guidelines are provided to help ensure the experience is positive.

The vocational GCSEs are designed to help increase flexibility in the compulsory examination system, in recognition of the fact that students have different learning strengths and consequently the heavily content-based science curriculum is not suited to some. The success of the courses will depend on the confidence placed in them by schools – we foresee that applied science GCSE will be appropriate for some schools but not others. The course may be more popular in post-16 colleges for students seeking an alternative to a traditional Double Award GCSE after having achieved only moderate grades.

At the time of writing, Ofsted released the first report on vocational GCSEs, finding achievement on Applied Science to be higher than that of students taking traditional GCSEs in some cases. Students and their parents liked the course, perhaps because of the perceived relevance and the more adult atmosphere created by encouraging links to industry and other colleges. The report suggested that professional development for teachers was needed to ensure further progress was made.

POST-16 DEVELOPMENTS

Post-16 courses began to change shortly after the introduction of the NC in the late 1980s. Prior to this, post-16 education in secondary schools comprised a fairly narrow range of A levels intended mainly for academic students, while those for whom A levels were not thought suitable or who did not wish to take them usually left school for FE colleges to take employment-related courses or for work-based apprenticeships. The advent of GCSE meant that access to academic-style post-16 education was widened, with a range of outcomes. These included:

- A levels being examined in more flexible ways, with terminal examinations being replaced by coursework and modular tests taken at regular intervals throughout the two-year courses;

- the range of subjects available increasing dramatically, creating competition for students and a gradual and severe decline in those willing to study science subjects; and
- curriculum developers working in science being challenged to create courses to meet these new circumstances.

At the same time, the government took the decision to reduce the number of syllabuses, now called 'specifications', from which teachers could choose. Each of the three examination boards was permitted to offer two different courses for each science subject. This means that now, in England and Wales, teachers can select from six alternatives. The aim of this decision was to enhance consistency in standard across syllabuses, as there had been concern that some courses were much 'harder' in terms of achieving high grades than others. In principle, examination boards now offer one 'standard' specification, offering A level sciences in a 'traditional' format, with one 'alternative' approach alongside. Teachers are free to choose between these. In this section we present two highly regarded 'alternative' projects established as part of the post-16 science curriculum and provide information about Vocational A levels or VCEs.

Context-led courses: the Salters projects

The Salters Institute has sponsored extensive science curriculum development mainly through the University of York Science Education Group and other partners. The name of the Institute is based originally on the wealth created by mining salt – 'salters' are the people responsible for mining and purifying this everyday chemical. In modern science education, 'Salters' is the name attached to a range of 'alternative' courses, all adopting a 'context-led' approach to teaching and learning science.

The first course with the Salters name was GCSE Chemistry, developed in the 1980s. This offered a range of units using themes in chemistry adopted from everyday life. Students studied each unit and in so doing learned the chemistry needed to understand the context. This principle was extended first to Salters Science for GCSE and then to three post-16 courses called Salters Advanced Chemistry (SAC, University of York Science Education Group, 2000a, b), Salters Horners Advanced Physics (SHAP, University of York Science Education Group, 2000c) and Salters Nuffield Advanced Biology (SNAB, University of York Science Education Group and the Nuffield Foundation, 2002). We will describe the basic principles of the Salters approach using SAC as an example.

Salters Advanced Chemistry (SAC) was the first post-16 context-led science course available in England, Wales and Northern Ireland. The course comprises 13 units divided between AS (five units) and A2 (eight units, see Table 4.3). Each unit has two main sections called the 'Storyline' and 'Chemical Ideas'. The Storylines provide the contexts for the chemical ideas. Only the chemical ideas required to understand each storyline are presented in each unit. This has one major consequence for learning: students are not presented with the entire A level specification content of a topic in one block of time. Traditionally, the topic chemical bonding, for example, would be taught in three – five weeks of teaching time, with any applications of the knowledge added towards the end. Within SAC, however, the topic is divided up, so that certain types of chemical bond are met in some units where the context permits while others feature in later units. The teaching is directed by the contextual information provided in the Storylines. From the perspective of learning chemistry, this means that topics are revisited throughout the two years of the course, so students' knowledge and understanding develops over this period. Also, each unit features a range of different chemical ideas, each of which is required to understand some aspect of the storyline.

Salters Horners Advanced Physics (SHAP) was the next of the three to be developed. This course has eleven context areas, six of which feature at AS level and five at A2 level. The Advanced Biology course is the most recent – the first trial was completed in June 2004. Titles of the units for all three courses are given in Table 4.3. The numbers of units in each course varies, but the principles described above are the same.

The unit titles are designed to appeal to young people and are written deliberately to explain how the specific aspect of SCIENCE applies in everyday life.

Comments and criticism of the Salters projects

Evaluations of the context-led approach are relatively rare. Barker and Millar (2000) report the findings of a two-year study investigating students' learning of basic chemical ideas. The study indicates that there are no apparent disadvantages of learning chemistry in this way, and that significant advantages appear to accrue in terms of students' understanding of certain key topics, such as chemical bonding and thermodynamics.

Anecdotal evidence suggests that most science teachers using the Salters approach find it very exciting to teach and beneficial in terms of

Table 4.3 Salters Advanced Science units

AS units	A2 units
Salters Advanced Chemistry	
The Elements of Life	What's in a Medicine?
Developing Fuels	Designer Polymers
From Minerals to Elements	Engineering Proteins
The Atmosphere	The Steel Story
The Polymer Revolution	Aspects of Agriculture
	Colour by Design
	The Oceans
	Medicines by Design
Salters Horners Advanced Physics	
The Sound of Music	Transport on Track
Technology in Space	The Medium is the Message
Higher Faster Stronger	Probing the Heart of the Matter
Good Enough to Eat	Reach for the Stars
Digging up the Past	Build or Bust?
Spare Part Surgery	
Salters Nuffield Advanced Biology	
Lifestyle, Health and Risk	On the Wild Side
Genes and Health	Infection, Immunity and Forensics
The Voice of the Genome	Run for your Life
Climate Change	Grey Matter
Plants: Can't Run, Can't Hide	

maintaining students' motivation for study. One of the authors taught SAC for a period in a sixth form college in a part of the country with low participation in post-16 education. SAC was well received, gaining very good examination results and stimulating a high proportion of students to study chemistry or chemistry-related subjects at university. This pattern has been reported widely among SAC teachers. SHAP and SNAB are at relatively earlier stages of their development, so evidence for their popularity and acceptance is more limited.

Some science teachers do not like the Salters approach. Their reasons include arguments based around the organisation of teaching science topics to quite a high level. Teachers with a traditional style prefer to work with topics as they themselves see fit, organising their programme of work to suit their students and school or college system. The Salters approach fractures the traditional teaching pattern into small pieces, providing a predetermined system that must be followed. Some teachers do not like the idea of being so closely directed. Other arguments centre on the impact on students.

Teachers argue that Salters courses require students to understand far more than is really necessary – the Storylines books are, themselves, filled with quite complex material, relatively little of which is 'formally' examined. Students who are not academically strong, it is argued, will not flourish through learning in the Salters style. Although the courses are not for the faint-hearted, Salters teachers would strongly counter this with the high levels of motivation seen among the majority of their students. A third argument relates to how the courses may be taught. Taking on a Salters course is a big challenge for teachers already established in a traditional way of teaching. In some cases teachers do not make the change easily, being still tempted to teach, say, chemical bonding, in much more detail when it first appears in SAC than the context demands. Also, the course materials include a wide range of activities, which if all were done, would take excessive amounts of time. Teachers at first find it hard to judge which are essential and optional.

However, the Salters projects have seen significant success and are now established as an alternative way of providing post-16 science education. Respect for the context-led approach has increased steadily, as no evidence has been found to indicate that this has a negative impact on students' learning and understanding, but does positively influence retention, develop positive scientific attitudes and retain motivation for further study.

Advancing Physics: presenting physics as cutting-edge and relevant

In response to the decline in the numbers of students pursuing A level physics, the Institute of Physics took the decision to invest considerable funding to support the development of an alternative post-16 course (Box 4.4). Directed by Jon Ogborn, the course includes modern physics with a strong bias towards applications and technology. Topics featuring at AS level are: Communication, Designer materials, Waves and quantum behaviour, and Space and time. Those at A2 are: Models and rules, Matter in extremes, Fields and fundamental particles of matter.

The progressive nature of the course is also seen in the course materials, which comprise books and supporting CD Rom. It is too early to know if the mixture of modern physics (quantum theory, cosmology, particle physics) and applied physics (sensing, communications technology,

BOX 4.4 ADVANCING PHYSICS

Advancing Physics is a contemporary course, developed by the Institute of Physics as part of a response to the falling numbers of students studying physics beyond 16.

Our core aims and objectives are that the course:

- makes physics exciting and relevant;
- is contemporary in content and modes of delivery;
- is attractive and accessible to the widest possible variety of students;
- sets physics in a variety of contexts, illustrating connections with everyday life, people, places and cultures;
- rewards students for initiative and commitment and allows them to develop their own interests;
- fully supports and recognises the use of essential mathematical methods in physics, helping students to understand them;
- fully supports teachers using extensive tried and tested resources and ongoing support.

Designed to attract students and give them a good basis for future decisions, the AS course offers a broad vision of physics as it is today. Providing an introduction to physics and its uses, it prepares the way for further study and focuses on the wide range of future careers for which physics is valuable. Teaching and assessment are designed to give students opportunities to pursue and develop their own interests and examines their understanding of physics.

Carefully balancing technological and applied approaches to physics in use, the A2 course deepens understanding of crucial ideas, giving students a wide-ranging and fundamental view of the nature of matter and the Universe. Mathematics in physics is further developed using modelling software. Teaching and assessment provides opportunities for personal student involvement and individual initiative.

(Edited from the Advancing Physics website, www.advancing physics.iop.org/, accessed 12 April 2004) computing/modelling motion), along with the flexibility to follow up interests in the coursework, will be enough to contribute to a reversal of 'the falling numbers of students studying physics beyond 16'. We have, however, certainly found enthusiasm among teachers and students working with the new course.

The topics included in Advancing Physics do not immediately match those of a more traditional physics course, but a close look at the contents reveals more similarities than the topic headings might initially suggest. More significantly, as a course devised under the auspices of the professional body and learned society for Physics in the UK, there is little doubt that Advancing Physics provides students with a taste of physics as part of SCIENCE in the twenty-first century.

Vocational A levels: Applied Science GCE

The UK education system has had a long history (see Chapter 2) of reliance on identification of 'academic ability', a feature that has taken, and is taking, a long time to change. This has been seen particularly in the reliance on A levels as indicators of likely academic performance at university and, in the case of science, a means of selecting specialists. In consequence, vocational qualifications have been widely perceived as the 'poor relation'. In recent years, work has focused on trying both to change the reliance on traditional academic selection (by whatever means) and to promote vocational qualifications as a realistic alternative of equivalent status to traditional A levels.

In England and Wales, one strand of the government's latest moves in this area has been to introduce 'Vocational' A levels, that is, courses that reach the same academic standard, but present knowledge in a much more work-related way than traditional A levels. The most recent version of this science qualification, called 'Applied Science GCE' is available from 2004, with the first examination in 2006. The course comprises 16 units, of which six are examined at AS level and the remaining ten at A2 level. The unit titles are given in Table 4.4.

The range of units is designed to provide wide choice. The course may be taken as a 'single' award or a 'double' award, that is, as equivalent to one AS/A2 A level or two. Students may decide to take two AS levels in Applied Science, but only one A2 level, one AS and one A2, or two AS levels and two A2 levels. To gain one AS level, students must study

Table 4.4 *GCE Applied Science units*

Applied Science GCE	
AS units	A2 units
Investigating science at work	Planning and carrying out a scientific investigation
Managing human activity	Medical physics
Finding out about substances	Sports science
Food science and technology	Physics of performance effects
Choosing and using materials	Controlling chemical processes
Synthesising organic and biochemical compounds	The actions and development of medicines
	Colour chemistry
	The healthy body
	The role of the Pathology Service
	Ecology, conservation and recycling

Source: AQA, GCE Applied Science Draft specification, 2004, available from the AQA website www.aqa.org.uk/, accessed 1 August 2004.

the first three AS units listed, while to gain a double award they must study all six.

The A2 year is a little more complicated. Although the investigation is compulsory, students must then choose from the remaining units. The specification gives rules as to which units may be chosen to ensure a balanced qualification is obtained. To gain one full A level, in the A2 year students must carry out the scientific investigation and then study two further units, while for a double award a further five units must be studied. As with the Vocational GCSE, some assessment will be by portfolio, allowing students to select work from their studies as they progress.

As the course title indicates, the aims for Applied Science are to ensure that students gain a realistic impression of how science is applied in industry and what career opportunities are available, as well as to understand basic principles in biological, chemical, physical and environmental science.

A glance at the content and the assessment demands suggests this is a high-standard course deserving respect. Close inspection reveals some similarities with the context-related courses described above – comparing

Tables 4.2, 4.3 and 4.4 shows a trend towards making science accessible through using unit titles designed to appeal. Allowing more freedom in assessment methods is also promising, as many students have disliked the emphasis on lengthy terminal examinations.

GCE Applied Science is, at the time of writing, the newest of the post-16 courses discussed here. The early versions were criticised for not being 'truly vocational' and for being taught by teachers with limited industrial experience. The new version is not, as yet, tried and tested, but remains a good prospect for enhancing further the range of science courses available to post-16 students.

CONSTRUCTIVISM: A LEARNING THEORY UNDERPINNING DEVELOPMENT

Here we have considered some of the more recent developments in science education. An astute reader could conclude that although these developments have aspects in common, they do not seem to be part of a coherent plan. As we mentioned before, science education is subject to a range of pressures, including political and social. Development of the subject is influenced by distinct, powerfully expressed viewpoints that sometimes pull science teachers in different directions.

Although we would like to offer a unifying framework for exploring science teaching and learning to help an inductee make sense of the variety of developments and initiatives, we think this would be rather ambitious. Nonetheless, it is useful for inductees to have a starting point for thinking about science learning – if only to help filter and prioritise some of the many initiatives likely to be met during a teaching career.

We suggest that the best starting point derives from what is often described as a 'constructivist' perspective on science teaching and learning (Bodner, 1986; Scott, 1987; Pope and Watts, 1988; Ross et al., 2000). Much has been written about the grounds for constructivism in science education, but we focus on what could be called the 'germ' of this perspective and how it might be useful as a basis for a science teacher's professional thinking. More detailed exploration is readily possible using the range of publications available, such as: Millar, 1989a; Fensham et al., 1994; and Taber and Watts, 1997.

The basic premise of the constructivist principle is that knowledge cannot be transferred wholesale from one mind to another. Instead, each of us has to build up, or construct our knowledge structures incrementally by anchoring new information onto existing material. A main thrust of the constructivist research effort in science education has been the exploration of students' ideas about aspects of science, partly to show how the construction process can take unexpectedly non-scientific routes, but also to catalogue the kinds of ideas students bring with them to their science lessons as starting points for knowledge construction.

The constructivist movement in science education has focused on students' ideas and beliefs, while other researchers have explored learning in terms of the actual mental apparatus involved. Johnstone (1989), for example, has studied how information is processed during problem solving. These approaches are analogous to studying chemistry at the macro- and microscopic levels – both are needed and there is a clear relationship between them (Taber, 2000d).

There is still a great deal more work to be done on how students' ideas develop with time, what factors channel and control this development and on understanding how learning science is contingent upon prior learning and the idiosyncrasies of the human brain. This provides the foundation for an on-going, progressive research programme exploring, for example, how students' ideas change over time and how learners consolidate and integrate their learning. Such projects would have much to contribute to teaching. Box 4.5 provides an example of a constructivist research project. The importance and relevance of research is a major focus of Part III of this book, so here we simply acknowledge that such developments in science education are influenced by research from within the science education community. Clearly, from what we have said about school science being socially-constructed, and subject to political pressures, research is only one of sources of impetus for change.

We think that constructivism is probably still the most useful starting point from which science teachers can think about how science learning occurs. A fuller account of this learning theory is given in Chapter 12. From its beginnings about 30 years ago as a way of exploring science learning, constructivism has become a complex framework for exploring teaching and learning that has a great deal to offer teachers. We have certainly benefited from its influence.

CONCLUDING COMMENTS

This chapter has reviewed a number of recent developments in science education at secondary level. Besides the KS3 National Strategy, we can

BOX 4.5 A CONSTRUCTIVIST RESEARCH PROJECT IN SCIENCE EDUCATION

The ECLIPSE Project: Exploring Conceptual Learning, Integration and Progression in Science Education

This project seeks to develop a better understanding of how conceptual learning occurs in science, with a particular focus on how well the learner's conceptions are integrated and how conceptual frameworks develop over time. The project is underpinned by a constructivist position on learning – that the nature of all learning is highly contingent (upon prior learning, learning context, features of language and so on). It is hoped that this project will help develop theory about learning science that will be of practical use to teachers, learners and those responsible for setting the science curriculum.

(From the project website at www.educ.cam.ac.uk/ eclipse/index.html, accessed 1 August)

see that the new courses fall broadly into two groups – vocational courses developed by government and context-led/modern style courses funded by professional bodies and charitable foundations.

Behind the UK Government's initiatives is the drive to raise standards of education at every level. Science is still perceived as a key subject in creating the nation's wealth and, as such, is promoted as a 'core' subject to the age of 16. However, as we have seen, the academic-style science curriculum has proved detrimental to the progress of a significant number of students, creating a decline in the numbers wishing to study science beyond the compulsory age. Coupled with this has been the drive to enhance the status of vocational courses in the UK. Hence, the new vocational KS4 and KS5 science courses have arisen. Adopting a work-related approach, these are designed to appeal to secondary schools as well as post-16 colleges. The assessment and content are also designed to be of comparable academic standard to traditional-style courses.

Private finance has largely been responsible for the development of context-led and other science education courses. The Salters projects, in particular, have adopted a specific stance on students' learning – that the content of an A level is best delivered through an everyday context appropriate to young people. The 'piecemeal' effect this has on teaching of traditional science concepts has been documented, but appears to have little, if any, negative impact on students' learning and some positive effects have been found.

The drive to continue to offer new courses and to improve those already available is continuing. Science educators have a strong reputation in the UK for being progressive in outlook and eager to create new ways of teaching. We ask inductees reading this material to give these new approaches careful consideration as they begin their teaching careers. Part of being a professional means exploring aspects of the subject that might seem uncomfortable or breaking with 'tradition', as these can lead to positive, if unexpected, outcomes. This, of course, is part of what Faraday meant in his comment about experimenting with which we opened this chapter.

THINKING ABOUT PRACTICE

- We state here that KS3 is regarded as a critical point for students — positive experiences here are likely to 'make or break' decisions about science taken later. Do you think the KS3 National Strategy for Science may aid or inhibit students' enthusiasm for the subject? How might science teaching benefit from the Strategy?
- Consider the advantages and disadvantages of context-led approaches to teaching science at KS5 level. Would you be prepared to teach a context-led course, or not? Explain your thinking.
- Consider the advantages and disadvantages of having vocational and academic science courses being taught alongside each other in a school. In what ways might students benefit? Might some 'lose out' in some way?

Chapter 5

What does a science teacher's expert knowledge look like?

An expert is someone who knows some of the worst mistakes that can be made in his subject and who manages to avoid them.

(Werner Heisenberg)

INTRODUCTION

In this chapter, we review some of the key ideas covered in this section of the book by asking what we might mean by 'expert subject knowledge' in the context of science teaching. As a subject specialist you will be expected to be a *science* 'expert', but clearly – in view of the broad nature of SCIENCE and the diversity of scientists we have discussed in Chapter 2 – you are *not* going to be at cutting-edge research level in all of the sciences. Your science expertise will be relative to your professional duties and responsibilities – you will be a science 'expert' as far as non-science colleagues and the students are concerned, if not an expert in any particular scientific field. Even if you enter the teaching profession fresh from top-level research in some 'sexy' area of SCIENCE, you have to accept and understand that you have chosen to let your status as a research scientist lapse to take on the new persona of educator. However, this does not in any way mean you should feel a sense of inferiority with peers who have continued to work in SCIENCE. As a teacher, you will not only develop new areas of expertise that these friends do not share, but you will also find you are acquiring a breadth of science knowledge that very few active scientists share.

Domains of expert knowledge for the teacher

There are three main areas of expert knowledge that contribute to effective teaching. The effective teacher knows the *individual* learners in the class. The importance of knowing your students cannot be over-emphasised and we shall have more to say on this topic later (see Chapter 6). Second, an effective teacher has a good knowledge of the science – here we focus on what we think 'good knowledge' is. Finally, the effective teacher has the know-how to teach science, which goes beyond just knowing the science. In other words, knowledge of science is not sufficient for a teacher – you must also have well developed knowledge of how to help *others* learn science. This is sometimes known as *pedagogic* knowledge. We will be returning to this theme later in the book. In this chapter, however, we start by considering the teacher's own understanding of science.

WHAT WOULD WE EXPECT AN INDUCTEE TO KNOW?

We start our consideration of this topic by outlining three key principles that apply to any new entrant to science teaching:

1 You don't know enough to teach science . . .
2 . . . but that does not matter . . .
3 . . . as long as you acknowledge this, and respond professionally.

Teaching is accepted as a profession requiring a high level of academic and professional preparation. In effect, teaching is now a graduate profession. To be considered a qualified teacher in a maintained school and, increasingly, in a post-16 college, a new entrant must complete successfully a recognised programme of professional training. To enter the programme an entrant is expected to hold a degree-level qualification. To teach science, an entrant to teaching must have a university degree or qualification recognised to be of the same standard, in a science or cognate subject.

So, to become a science teacher you must be a science graduate. Indeed, many programmes of graduate training normally expect applicants to hold a first- or second-class honours degree, or equivalent. This means that the entrant has studied a science subject at higher education level, and has demonstrated a proficiency in that subject to a high standard in university level examinations. We thoroughly approve of this expectation that secondary and college science teachers should normally be expected to hold 'good' science degrees. Science is a complex and subtle business, and without that high level of personal knowledge and experience it seems unlikely that a teacher can do the subject justice. However, we see this as a necessary rather than a sufficient condition – it certainly does not guarantee sufficient subject knowledge for effective teaching.

By having a science degree, the entrant will have a broad and deep knowledge of important aspects of a disciplinary area (of what Kuhn called the 'disciplinary matrix', see p. 26). The inductee knows about the important theoretical frameworks that are the key explanatory devices in the subject area, about the standard methods used in the field, what counts as scientific evidence and how it is analysed, assessed and explained. The entrant may also have some knowledge about how the field has developed historically. Some new entrants bring additional experience of research in an academic or industrial context, or of applying the science in the workplace.

Yet, we suggest, there are at least two strong reasons to see this background as necessary – but hardly sufficient. The first relates to the breadth of an inductee's science knowledge and the second is what we refer to as the 'robustness' of the knowledge.

The breadth of knowledge needed to teach science

The first reservation should be obvious to the reader in view of our discussion of 'SCIENCE' at the beginning of the book (see Chapter 1). Science graduates are hardly a homogeneous group. Degrees in physics, chemistry, botany, genetics, metallurgy and so on, each provide detailed subject knowledge about a particular field within SCIENCE. Some science degrees may provide a useful 'applied' aspect, but may be even more specialised, for example, pharmacy, marine conservation and forestry. Engineering degrees will often be biased towards maths and will normally include large areas of physics, but do not usually provide a broad science background. Different engineering courses such as mechanical, electronic, civil and chemical may themselves give very different coverage, while some may be highly specialised, such as mining engineering or, like production engineering, have significant foci away from SCIENCE.

So, someone with a degree in a science or related subject will at best only have studied *some* science in depth. When their 'area of expertise' is compared with the school science curriculum, large areas of knowledge to be taught will fall outside this area. If science teachers only taught within their specialist areas this would not be a problem. However, this is not usually the case. Although some college teachers may limit their teaching to one main science subject, in most secondary schools all science teachers contribute across the science curriculum at lower secondary (KS3, 11–14-year-olds) and often beyond. This approach has at least tacit official approval in the English education system, where 'biology', 'chemistry' and 'physics' are no longer officially part of the mandatory curriculum, being replaced by 'science'.

Training to be a science teacher in England therefore means training to *teach science*, not to teach *a* science (see Table 2.1, p. 22, and Box 5.1). That is not to say that trainee science teachers do not usually have a specialism within science, nor to deny that *many* teachers in schools do not have a timetable biased towards their specialist subject. The extent to which individual science teachers specialise within science varies considerably. Factors such as school traditions and philosophy, science department/faculty policy, personal preference and recruitment issues influence the actual aspects of science that a science teacher may teach.

School science as a politically demarcated entity

This policy means that training to become a qualified teacher of, for example, biology alone rather than science, in the state system is not possible. This is a particular issue for some graduates who might feel that their interests and strengths best suit them to offer specific subjects. A particular case would be some graduates in physics or engineering who might feel that their interests and strengths best suit them to offer physics and maths as their teaching subjects. (It is worth noting that the description of the Advancing Physics course in Box 4.4 on p. 85 makes two references to the intimate relationship between mathematics and physics.) Physics and maths are often found in combination in degree courses, so it is also possible that graduates may wish to teach both. There are undoubtedly many teachers in post who do teach this combination, but the new entrant must decide to train to teach either maths, effectively excluding physics for at least the ITT year, or to train as a science teacher prepared to teach chemistry and biology as well as physics.

The UK Government does not make the policy on subject teaching obvious. The Teacher Training Agency (TTA) website (www.useyour headteach.gov.uk, accessed 2 April 2004) advises interested potential

BOX 5.1 THE STRUCTURE OF THE SECONDARY CURRICULUM

The National Curriculum (NC) in England and Wales must be followed by maintained schools.

Foundation subjects must be taught. These are English, Mathematics, Science, Design and technology, Information and communication technology, History (KS1–3), Geography (KS1–3), Modern foreign languages (KS3–4), Art and design (KS1–3), Music (KS1–3), Physical Education and citizenship (KS3–4). Of these, three foundation subjects are designated as being the core subjects in the curriculum: English, Maths and Science.

The NC sets out 'Programmes of Study' (PoS), 'Breadth of Study', and 'Attainment Targets' (AT).

> The programmes of study set out what pupils should be taught in science at Key Stages 1, 2, 3 and 4 and provide the basis for planning schemes of work. ... The knowledge, skills and understanding in each programme of study identify the four areas of science that pupils study:
>
> Sc1 Scientific enquiry
> Sc2 Life processes and living things
> Sc3 Materials and their properties
> Sc4 Physical processes.
>
> (DfES/QCA, 1999: 6)

Sc1 is about the processes of science (see Chapter 2); Sc2 is basically biology; Sc3 is mainly chemistry (with some earth science); Sc4 is mainly physics with some astronomy topics:

> The breadth of study identifies contexts in which science should be taught, makes clear that technological applications should be studied, and identifies what should be taught about communication and health and safety in science.
>
> (DfES/QCA, 1999: 6)

> The attainment targets for science set out the 'knowledge, skills and understanding that pupils of different abilities and maturities are expected to have by the end of each key stage' ... in the four main sections of the programme of study: scientific enquiry, life processes and living things, materials and their properties and physical processes.
>
> (DfES/QCA, 1999: 7)

teachers to explore subject combinations such as physics and mathematics, implying this is a viable pairing. Indeed, visitors to the site are told that: 'At secondary [level], teachers may teach *one or more* of the National Curriculum subjects, or one of an increasing number of vocational subjects on offer' (emphasis added). The site does not appear to distinguish between 'teach' and 'train to teach', as technically it is true that teachers often end up teaching more than one NC subject. However, anyone wishing to take their interest further and considering applying is redirected to the website (www.gttr.ac.uk/, accessed 2 April 2004), of the Graduate Teacher Training Registry (GTTR). Reading GTTR material clarifies that there are courses for those wishing to specialise in mathematics *or* physics, but no courses offering *training* to teach both. This policy presumably derives from a belief that science and mathematics are distinct teaching subjects, each having its own pedagogy and each needing specialist-trained teachers. We suspect that this inhibits a number of potentially good candidates from entering teacher training, although in part this separation of maths and physics reflects a strong government position on the importance of science and mathematics per se as core subjects. We welcome this recognition of science in the curriculum, but suggest that enabling dual training in maths and physics would be sensible in the face of an on-going shortage of science teachers, especially physical science specialists, and maths teachers.

We accept that differences in training needed to become an effective physics teacher and an effective maths teacher are apparent, but there will also clearly be many similarities. The UK Government may consider that it is 'too hard' for graduates to train to teach two different subjects in a one-year ITT course. However, we note that trainees are considered able to train to teach both English and drama in one year: subjects that we suspect have a similar relationship to maths and physics.

The construction of the entity labelled 'school science' may mean that there is a different agenda here, perhaps a need to give the broader subject an accepted and important entity. After all, we argue, the practice of the discipline of physics probably has at least as much in common with that of mathematics as it does with the practice of many of the life sciences. In other words, we see the removal of the subject labels of biology, chemistry and physics from the official secondary school curriculum, the training requirements and restrictions on subject combinations as part of a policy to establish the notion of science as a subject within the education system.

Debates about demarcation, e.g. distinguishing SCIENCE from 'non-science', have a long history in academic debate (e.g. Popper, 1979), but we are more concerned about the practical implications of the official demarcation of school subjects. We are concerned that once 'science' is established as subject, rather than the specialist subjects, schools need only recruit and retain 'science teachers' who have been trained to teach 'science'. We believe that for many 11–16 schools in England this is reality. Schools with a full complement of science teachers may have no physics or chemistry specialists and schools looking to appoint new science teachers in many parts of the country are lucky if they entice any physical science specialists to apply.

We are not suggesting that the purpose of promoting 'science' as the recognised school subject is simply an official policy designed to help disguise a severe shortage of physical science teachers: but this is clearly one likely effect. It is important then to ask whether this matters. After all, we have already strongly argued that the inductee must metamorphose from physicist, engineer or zoologist or whatever former role to take on the new professional identity of science teacher. Does it not follow that a subject specialism *within* SCIENCE is just part of the old professional identity that this book claims the science teacher must leave behind? We argue quite the opposite. If the entity of school science is to effectively provide an educational basis for reflecting SCIENCE, then this has to be informed by the full range of scientific disciplines. One area where this is especially important is in the nature of the science teacher's subject knowledge.

Put simply, no new science teacher is likely to have detailed background knowledge of all areas of science in the curriculum. Thus, if school science is to reflect SCIENCE in a representative way, then the professional community of science teachers should *collectively* have subject knowledge backgrounds reflecting the range of knowledge and practices in the SCIENCES. The Induction Standards for NQTs talk about teachers identifying areas where they need to improve their professional knowledge, and responding to these needs *with support*.

Teachers are required to continue to meet the requirements of the Knowledge and Understanding section of the Standards for the Award of QTS, and build on these. Specifically, they show a commitment to their professional development by:

- identifying areas in which they need to improve their professional knowledge, understanding and practice in order to teach more effectively in their current post, and
- with support, taking steps to address these needs.

(TTA, 2003b)

Clearly, some departments may not be in a position to offer much support in helping inductees develop their subject knowledge in areas of the science curriculum if there are no experienced specialists in, say, physics and chemistry, working in the school.

The concern that we have expressed above about the lack of physical scientists teaching in some schools, along with the way the 'science' label could disguise this shortage is not merely based on our personal views. The RSC's Chemical Education Research Group has explored the potential shortfall in chemistry specialists entering teaching (Doherty *et al.*, 2000). Perhaps even more tellingly, the IoP has taken the view that the lack of science teachers with strengths in physics probably means that there are many schools where KS3 physics (i.e. Sc4, see Table 2.1, pp. 22–3) topics are being taught by teachers who are neither knowledgeable, nor

BOX 5.2 TEACHING ACROSS THE WHOLE OF SCIENCE

The government is aware that teaching beyond one's 'comfort zone' can be a challenge, and may reduce the effect to which teachers (and departments) can face other demands. The following statement is taken from the Key Stage 3 Strategy (see Chapter 4) publication *Framework for teaching science: years 7, 8 and 9*:

the degree to which a science department adjusts its current practice [should be influenced by] . . . the extent to which the department is staffed by teachers, including those qualified in physics, chemistry, biology and other science qualifications [*sic*] such as geology or biochemistry, who may need support with their planning and teaching across the whole of science, especially when they are teaching beyond their specialism.

(DfES, 2002: 9)

confident, in the physics. The IoP's concern has been sufficient to cause them to invest in a major project – Supporting Physics Teaching 11–14 – to provide training and support materials for these teachers (see Box 5.3) – 'to help teachers develop a better understanding of physics'. We will have more to say about the importance of in-service education, or 'continuing professional development' (CPD) on pp. 215–16.

BOX 5.3 SUPPORTING PHYSICS TEACHING 11–14

This new initiative from the Institute of Physics aims *to support non-specialist teachers who teach physics in the later years of primary education and the early years of secondary education.* The project will develop professional development programmes supported by a set of CD ROMs. The aim is to help teachers develop a better understanding of physics, allowing them to experience for themselves something of its fascination and to develop greater confidence in their teaching of it.

(From the IoP website, www.education.iop.org/
Schools/SPT/, accessed 5 April 2004, emphasis added)

So, our first concern is with the breadth of knowledge acquired during a typical science degree. Of course, some new entrants hold degrees covering broad areas of SCIENCE rather than being trained within a particular scientific discipline. In one sense, such degrees may offer a much better background for science teaching. However, if school science is to reflect the broad range of traditions within the SCIENCES as discussed above, then we would not necessarily wish to see too many scientific generalists among the profession.

Robust subject knowledge and teaching science

A second limitation of a new entrant's subject knowledge is what we refer to as its 'robustness'. At first sight, we seem to either know something or we don't. However, this is a gross over-simplification, as science teachers soon appreciate from their own experiences with students. Knowledge is not an 'all-or-nothing' entity that we may or may not possess. A moment's reflection reveals that things we 'know' may be more or less readily brought to mind. In other words, just because something is 'stored' in our memories, this does not mean that memory is easily accessed. We have all experienced that so-called tip-of-the-tongue phenomenon when we are unable to recall some specific piece of information that we are sure we do know. Memory studies indicate that memories may be more or less readily accessed, or 'brought to mind' and that recall is often context-dependent. A good example is when you do not immediately recognise someone you know when you see him/her in a place you would not have expected to meet.

Having knowledge is not an end in itself, but to make it valuable, we need to use knowledge in various ways – and some ways of using knowledge may be more demanding than others. One especially taxing example is learning and using a new language. A friend of one of the authors who is a Spanish native speaker and close to fluent in English, is struggling to learn Norwegian as a third language. She describes her experience of going into a McDonald's in Norway to order a burger:

> You know, I rehearse what I want to say in Norwegian while I'm standing there in the queue. I feel my knees shaking and I sweat. I hear everyone in front of me giving orders so easily. Then it is my turn. My mouth opens. No sound comes out. Then I say, 'A Big Mac, fries and coffee, please!' It's so embarrassing!

Science teachers may experience similar nerves in learning and teaching the language of a new aspect of SCIENCE. The unfamiliarity of the material and the degree of 'foreignness' can lead to sweating and shaking knees; a feeling that things are not quite 'under control'; that all sorts of tricky and unanswerable questions may arise; that other colleagues more competent in this area are certain to criticise. Overcoming these feelings takes time as well as experience and is something that is of particular importance in teaching. There is a well-known aphorism that is a common part of teachers' craft lore. This is the idea that 'you don't really understand something until you have to teach it'. This bland statement reflects a common experience of teachers, who find that the type of intimate knowledge of a subject that equips one to teach it effectively is a much more thorough and *intelligent* type of knowing than one typically obtains as a student of the subject. Taking the language learning analogy further reveals that learning a foreign language thoroughly takes most people long hours of dedicated study – and even then their useable vocabulary may be only a fraction of what a native speaker knows.

We can appreciate this distinction if we think of how we learn for a purpose. When we study a subject in an academic context we are often focused on how our learning will be assessed. Typically, we learn the academic subject in a form useful for passing an examination: an examination with questions we can normally predict within certain parameters. At degree level, guidance in terms of a syllabus, lecture notes, perhaps textbooks written by our lecturers and past papers, is made available. We know the type of questions, the range of topics and often even how the questioner would answer the question themselves. We also know that we will probably be able to avoid topics outside our preferences and that in our chosen questions we can afford to omit some parts of the question, or even get the answers wrong and still be awarded our good degree.

Consider instead a situation where you have to cover all the set topics and be able to answer all sorts of questions, without notice, from a group of 25 or 30 'examiners'. Furthermore, these examiners do not think, or even realise that *their* questions should be circumscribed by *your* perception of what is relevant to the topic. Nor can you assume that there is sufficient common ground to take the relevant background knowledge for granted in your explanation. The experience of the teacher in the classroom is far more demanding than that of the student in an assessment, or the author's friend in a Norwegian McDonald's. Teachers cannot afford to omit, or get wrong, very much material if they hope to 'pass' the examination set by their learners. Students must have confidence that the teacher is an *authority* on the subject if they are to invest time and effort in trying to learn.

THE SCIENCE TEACHER AND ALTERNATIVE CONCEPTIONS

Shortly, we will seek to reassure the reader that most inductees into science teaching, however patchy their science knowledge on starting training, soon develop a sufficient breadth to their subject knowledge to be able to operate effectively in the classroom. Before we move onto this reassurance though, we do need to highlight one important and pertinent aspect of science knowledge that is fundamentally important to science teachers.

A learner's knowledge of a subject can best be characterised in terms of which aspects of the subject are known and understood. One might imagine that the alternatives here would be 'knowledge' and 'ignorance', or – at least – that the learners' knowledge should lie somewhere on the ignorance-knowledge dimension. In practice, learners' science knowledge transpires to be more complex than this. Research shows that learners often come to science lessons with *alternative* conceptions (also called misconceptions) of key science topics. The teacher's job here is not to replace ignorance with knowledge, but to help learners move their existing ideas towards the more orthodox understanding represented in the curriculum. As might be imagined, this is often an even more demanding task than teaching something to someone who begins by knowing nothing of the topic.

The topic of learners' alternative conceptions in science is well documented (for example, Driver *et al.*, 1994; Barker, 1994; Barker and Millar, 2000), and includes a good deal of contention (even the choice of the term 'alternative conception' would be seen as contentious to some experts in this field). This volume cannot do justice to the vast literature available, but the significance of this topic for science teachers is such that we will be providing some essential background and directing readers to explore the topic further for themselves.

At this point, we draw on one example from the topic of force and motion. According to many students, 'if a body is moving there is a force acting upon it in the direction of the movement [and] if a body is not moving there is no force acting upon it' (Watts, 1983: 226). See Box 5.4 to ensure you understand what is wrong with this statement. This idea seems to be very common, and has been associated with the historical notion of 'impetus'. Gilbert and Zylbersztajn (1985) found that 85 per cent of a sample of 14-year-old UK students 'associated force and motion' in this way. When asked about simple situations involving projectiles, 'they saw the stone as having a force upward away from

---

BOX 5.4 NATIONAL CURRICULUM STATEMENT ON FORCE AND MOTION

Pupils should be taught: that unbalanced forces change the speed or direction of movement of objects and that balanced forces produce no change in the movement of an object.

(NC, KS3, Sc4, 2c)

---

the person's hand as the stone moved upwards; the cannon ball was seen to have a force away from the cannon, moving it through the air' (1985: 115).

This type of thinking has been described as 'intuitive physics' and Gilbert and Zylbersztajn argue that 'for most children' the Newtonian model for relating force and motion (as taught in schools) 'is anti-intuitive' (p. 118). Other workers support this viewpoint. For example, Driver refers to a student who 'seems to consider the natural state of any object to be the stationary state, and when the initial impetus given to the ball is used up it returns to this state' (1983: 26). Another researcher comments:

> Recent studies . . . indicate that many people have striking misconceptions about the motion of objects in apparently simple circumstances. The misconceptions appear to be grounded in a systematic, intuitive theory of motion that is inconsistent with fundamental principles of Newtonian mechanics.
>
> (McCloskey, 1983: 114)

Learners are creative in interpreting situations to fit their intuitive physics. Laurence Viennot, a French researcher, reports that when there is no force acting in the direction of motion learners will often invent one and, similarly, if there is no motion in the apparent direction of the applied force, a learner may invent a force to produce a zero resultant (1985: 434).

In a US study, McCloskey found that over 90 per cent of high school students tested demonstrated an impetus theory of motion before they were taught Newtonian physics, and '80 per cent of the students retained this belief *even after finishing the course*' (1983: 122, emphasis added). Considering the research evidence available, Gilbert and Zylbersztajn concluded:

> The studies [reviewed] support the view that pre-Galilean ideas about force and movement are not only prevalent among school children, but also *in certain cases do persist even after years of formal exposure to physics teaching.*
>
> (Gilbert and Zylbersztajn, 1985: 117, emphasis added)

Students appearing in our classes can neither be assumed to have understood and learnt all the science they have met in earlier lessons, nor to be like 'blank slates' with no preconceived ideas. One complication of students' alternative conceptions research indicates that the student often thinks that they understand the science, so do not realise their error. A student who thinks they already understand the key principles in a topic, as opposed to one who knows nothing, is not able to help the teacher by pointing out where the 'gaps' in their knowledge are, because there are no 'gaps', just ideas that mismatch with the science to be learnt. Responding to learners' ideas in science is a major challenge for the science teacher. To borrow and develop a common aphorism, a square peg only seems out of place in a round hole once you have established the shape of the hole – so the alternative conception may not seem 'out of place' until the learner's knowledge of the rest of the topic closely matches the accepted science.

The observant reader may have noticed how in the discussion of intuitive physics above we have talked about 'learners' rather than students. This was a deliberate choice of word, as some common alternative conceptions of science topics have been found among college students, university students and graduates. The research shows that even trainee teachers are not immune. This is actually very significant for the induction of new science teachers. Institutions offering courses leading to QTS must:

> ensure that trainee teachers' achievement against the QTS Standards is regularly and accurately assessed, and confirm that all trainee teachers have been assessed against and have met all the Standards before being recommended for the award of Qualified Teacher Status.
> (TTA, 2003a: 16)

In view of the training standard which requires teachers to 'have a secure knowledge and understanding of the subject(s) they are trained to teach', this could be taken to mean that all trainee teachers have to be tested on their knowledge and understanding of all the science in the NC. In practice, a PGCE course is divided between 12 weeks of HEI-based work and 24 weeks of school-based work. Any comprehensive programme for teaching and assessing subject knowledge across the entire science curriculum would take up a large part of that limited time. In practice, most HEIs rely on a self-audit process. A trainee teacher is provided with a copy of the curriculum and is asked to 'audit' their subject knowledge in terms of their strengths and relative weaknesses.

105

This audit then forms the basis of a personal action plan for developing subject knowledge in areas of weakness.

The onus is on the trainee to identify and address relative areas of weakness and the training institution takes responsibility for ensuring *the process* has been carried out. In terms of the current training regime, there is little realistic option to this. Institutions may offer help with directing trainees to sources of support and may devise peer tutoring and/or limited subject knowledge tuition, but it would be totally unrealistic to examine the trainee's subject knowledge in any detailed and systematic way.

From what we have said above, trainees may well have their own alternative conceptions of some aspects of science in the curriculum. These conceptions will not seem 'alternative' to the trainee, so are unlikely to be identified through a simple audit process. There is certainly increasing material available to help trainees test themselves and obtain feedback on their level of knowledge and understanding. However, without being able to identify the topics in which they have alternative conceptions, the trainees would need to test themselves across the curriculum: and this time-consuming process would compete with the many other demands on a trainee teacher's time.

Clearly, many alternative conceptions inductees hold are likely to be exposed and challenged through conversations with tutors, mentors and peers, reading and lesson preparation and observations of more experienced colleagues. However, some alternative conceptions will survive this process and become part of the teachers' store of science knowledge used in their teaching. In this way, some alternative conceptions may be passed to students, and perhaps even to another generation of teachers in due course. If this seems fanciful, see the discussion on ionic bonding in Chapter 2, and the discussion of 'pedagogic learning impediments' in Chapter 7.

For example, one of the present authors remembers a conversation with his Head of Chemistry in his first post, who clearly found it incredible that the new, qualified teacher in the department did not realise that strong acids always had a pH of 1. The author was just as incredulous that his Head of Department could hold such a belief. A lunchtime meeting involving universal indicator strips, spotting tiles and measuring cylinders was agreed to settle the issue. Serial dilution demonstrated that even those teaching for years and very confident in their understanding of their degree subject can still hold alternative conceptions that seem unsupportable to other teachers.

This scenario concerns two graduate chemists and qualified science teachers disagreeing over the chemistry turning to empirical investigation to see who was right. This would seem to be a perfectly sensible and professional way of dealing with the question. Two professional colleagues find they disagree over a matter of substance (*sic*), so they turn to a higher authority. This could be seen as part of the process of continuing professional development that the government requires from teachers and which we would wholeheartedly endorse. Yet in terms of the absolute nature of the wording of the QTS standards, such a scenario should no longer happen: teachers are supposed to have sound subject knowledge *before* they are designated as 'qualified'.

Reality and common sense tell us that graduates who are found to have weak subject knowledge, and seem unable to respond to the problem, should not be allowed to qualify as teachers. However, in view of the breadth of the science curriculum, the more narrow nature of undergraduate courses and the ubiquitous nature of common alternative conceptions in science, it seems inevitable that all inductees attaining QTS will still have flaws in their science knowledge. A healthy attitude would be to encourage all teachers, especially those recently qualified, to explore their subject knowledge with colleagues to expose these. Some commentators have suggested that the way the training standards are worded is likely to discourage teachers from admitting that they may not have a totally secure subject knowledge across all the topics they might teach (Goodwin, 2001). The danger of such discouragement is that teachers hold, teach and so reinforce their own alternative conceptions, while passing them on to the students they are educating.

THE SCIENCE TEACHER'S SUBJECT KNOWLEDGE

At this point the professional work of the teacher – faced with 30 or more inquisitors with open agenda – may be starting to sound like some kind of medieval torture of 'trial-by-teaching'. It is worth reminding readers that very early in their professional careers most science teachers do develop the type of subject knowledge that enables them to undertake, and pass, frequent classroom 'examinations' without losing many marks at all. The science knowledge of the *effective* teacher is broad enough to permit teaching across the full range of science topics represented in the curriculum. It also becomes robust enough so that s/he can – at virtually no notice – appropriately respond to questions about any aspect of school science, no matter how apparently oblique or idiosyncratic the question. This means that any new entrant to science teaching is required to look at the whole science curriculum and read about areas where knowledge is lacking. For some trainees this may be quite a substantial undertaking such as learning all the physics in the NC. Besides this, SCIENCE is a dynamic aspect of our culture, so even experienced science teachers need to continue to read about the subject, reading science articles in the quality papers, magazines such as *New Scientist* and/or reading some 'popular science' books.

Achieving breadth of knowledge is only part of the task, for the science has to be understood well enough for the teacher to be confident enough to face a class. This type of knowledge development, making your knowledge suitably robust for teaching, comes about not by 'reading around' the subject, but through detailed 'preparation for teaching'. We will say more about preparing for teaching in Part II. Here, we comment simply that knowing the science is not enough to be able to teach it well. The teacher does not just recite or regurgitate facts but, rather, transforms their own level of knowledge into lesson episodes to help students learn at their own levels. Presenting science in a way that will interest and be sensible for 11–18-year-olds with very different existing levels of knowledge and understanding is, itself, a major and complex domain of knowledge. Earlier, we referred to this as *pedagogic* knowledge and the effective teacher uses both their subject knowledge and their pedagogic knowledge, alongside their knowledge of the students, to make professional decisions about how to teach a particular class of learners.

THINKING ABOUT PRACTICE

■ Use Table 2.1 on pp. 22–3, which outlines the Science content of the National Curriculum as a basis for a 'science knowledge audit' for yourself. In which areas would you need to review knowledge due to weaknesses and in which do you consider yourself to be strong?

■ How robust is your knowledge? Try these questions in each of physics, chemistry and biology:

Physics: Imagine an object moving, perhaps passing in front of you from left to right. Can you deduce from the information given whether there is a force acting on the object and, if so, the direction of the force?

Chemistry: Explain why if you blow on a match it goes out, but if you blow on a fire it will burn seemingly more strongly. Where does the energy released during fuel combustion come from?

Biology: Why are large animals scarce?

■ The alternative conceptions research discussed here was largely undertaken over twenty years ago, before the NC was introduced, and when many primary schools taught minimal science. Do you think that a replication study is likely to find similar results today?

Part II

Teaching science in the secondary school

In Part I we explored the notion of science as a school subject, and we made a distinction between SCIENCE (the professional activity of scientists), the distinct science disciplines (such as BIOLOGY, CHEMISTRY and PHYSICS), and school science – the subject labelled 'science' in the curriculum. We made it clear that scientists have a variety of disciplinary backgrounds, but are expected to take on the professional identity of 'science teacher' in which they are expected to have expertise within the secondary school context across all aspects of school science. We consider new entrants and early career teachers as inductees, people in the process of being inducted into the profession. According to the UK Government this involves demonstrating competence in meeting a range of teaching standards (TTA, 2003a), but we would argue there is something more fundamental involved: accepting and understanding what it is to be a professional science teacher. In Part I we considered one key aspect of this – developing subject knowledge in moving from the deep but narrow specialist knowledge of a scientist to the broad background needed to teach science as a school subject.

Part II moves the reader on from considering the nature of the subject, to engaging with how a teacher has to *transform the subject* in order to teach effectively.

We will define effective teaching very simply: *as facilitating appropriate learning in the students*. This rather bland definition has three key elements. First, and perhaps rather obviously, we do not consider teaching to have occurred if no learning has taken place. This rather simple point causes difficulties for many inductees who think they have explained a topic wonderfully, so the students' inabilities to make sense of it must derive from their lack of intelligence, motivation or attention, not the inductee's failure to explain appropriately. A moment's thought suggests that inductees at this stage may have good subject knowledge, but lack basic pedagogic knowledge.

The second element of our definition of effective teaching is the 'appropriate' aspect: *what* is learned must relate to what the teacher is supposed to teach. There is a process through which the teacher is guided: from curriculum to 'schemes of work' to lesson plans to classroom action. In practice, this process is never quite as easy: especially when we remember that we have emphasised the teacher as being a professional. In this part we will look at this process by which the teacher judges appropriate learning.

The third key aspect of the definition is the plural nature of 'the students'. A class of 25 or 30 students comprises individuals, however the group may be put together, with different strengths and weaknesses, aptitudes and enthusiasms. Another common mistake for some inductees is to judge their success in the classroom by the apparent learning of the few most able or responsive students, rather than to consider that a lesson should be successful for each student present if the lesson is truly effective.

This part, then, explores key themes that teachers need to consider in converting their subject knowledge into professional classroom expertise. These will include the nature and role of practical work, students' thinking about science, and the use of language, models, analogies and metaphors in teaching. The section will draw heavily on key ideas about learning that inform effective pedagogy.

Chapter 6 will set out to encourage a thought process that takes the prescribed content and considers the constraints imposed by features of (a) the specific learners; (b) the learning process; (c) the teaching processes.

Chapter 7 takes up from the previous chapter by looking at planning to teach the science. In other words, where Chapter 6 considers issues that would apply in principle to teaching any material, Chapter 7 explores how the nature *of the subject* should influence the way in which the principles of planning are carried through in planning *science* lessons.

Becoming a teacher implies taking on a new 'role', and different teachers can conceptualise the teacher's role very differently (Tobin *et al.*, 1990). The way an inductee conceptualises his/her role as a teacher is likely to set the pattern for teacher behaviour and is likely to continue for many years. In Chapters 8 and 9 the teacher's role will be presented as the leader in a process of joint construction of knowledge.

The metaphor of learning to be an 'actor' is used in Chapter 8 to introduce key elements of the process involved in learning to stand in front of a critical audience. Chapter 9 explores aspects of the pedagogical knowledge required to become an effective science teacher. This builds on material presented earlier, but focuses on how to present science in order that students may learn what the teacher intends.

The final chapter in this part, Chapter 10, considers how the teacher judges whether their teaching has been successful – that is, we look at evaluation of teaching. However good our planning has been, there is always something to be learned for the future – developing effective teaching depends on a cyclical process. With lessons being complex events, this means learning from successes as well as mistakes. In this chapter we also introduce aspects of assessment, as this helps to judge how successful our teaching has been.

Chapter 6

Planning to teach the curriculum

That is the difference between good teachers and great teachers: good teachers make the best of a pupil's means: great teachers foresee a pupil's ends.

(Maria Callas)

INTRODUCTION: PREPARING TO TEACH

We have pointed out that an effective teacher draws upon an expert subject knowledge base. This knowledge is different, both in scope and nature, to the knowledge that a new science graduate is likely to have acquired during their academic training. Clearly, part of the process of taking on the new professional identity of a science teacher concerns acquiring the type of subject knowledge that acts as the foundations for teaching science in school and college.

However, a second suggestion was that to some extent the robustness of that expert knowledge derives from the process of 'preparing to teach'. In this and the following chapter we will consider what is actually involved in preparing to teach, to help you think about aspects of planning your teaching.

Our third claim was that subject knowledge is only one important domain of knowledge needed for effective teaching. Teachers also draw upon their knowledge of individual classes and on something called 'pedagogic' knowledge. Put simply, pedagogic knowledge is knowledge about how to help someone learn something. Preparing to teach requires the teacher to draw upon these three different domains of knowledge. To teach science effectively to young people you have to know the science, know the students and know how to facilitate learning; then you can plan how you will *transform* the science to produce a lesson that will help the particular group of students learn about it. This is not a linear process – rather, preparation for teaching has a dialectic quality: many factors have to be weighed and considered at once. Later, we will suggest that once you start teaching you will appreciate that this dialectic extends beyond the preparation stage (see Part III).

We will begin by looking at the curriculum, go on to consider schemes of work (SoWs), exploring progression and continuity and offer principles to inform lesson planning.

WHAT IS 'CURRICULUM'?

The term 'curriculum' relates to what is to be learnt in school. The formal curriculum sets out what 'should' be taught, for example, the NC in England (available online at www.nc.uk.net/index.html, accessed 1 August 2004). The formal curriculum can be found in various curriculum documents such as school policies and schemes of work. The term 'hidden curriculum' is sometimes used to refer to things that may be learnt that are not part of the formal curriculum. Examples include messages that the school sends unintentionally: that academic learning is more important than vocational learning, that passing the examination is what matters, that less able students are less valued or that girls are not suited to certain careers. The formal curriculum, however, is the starting point for planning.

The curriculum: the starting point for planning

One of the authors recalls his first teaching job:

> My first class on a Monday morning was a mixed ability group of 14-year-olds who were timetabled to be taught their one hour of physics for the week. The experienced head of department informed me that for the first six weeks of the term that year group studied 'kinetic theory' and indicated a suitable textbook that might be useful. Perhaps there was actually more structure than this recollection suggests! I was directed to a science topic and left to exercise professional judgement from that point on.

And the other author recalls hers:

> I shared the chemistry teaching with one other teacher. At the end of my first year in the post, we decided to re-write the 3rd year (year 9) course. There was no 'formal' guidance. We decided to look at the examination syllabuses used in the 4th and 5th year and work backwards to think about what the students would need as good preparation for taking chemistry (then an optional subject) as an exam subject. We wrote the course ourselves using books and our own ideas.

These episodes occurred in the early–mid-1980s. At that time, examination classes followed examination syllabuses (called 'specifications' today), but schools had considerable autonomy in deciding what and how to teach years 7–9. In practice, the school would usually have some sort of plan designed to act as suitable background for examination work in the upper school (14–16-year-olds, years 10 and 11). The science taught in primary schools was extremely varied, with few primary teachers being qualified in, or confident in, teaching science. Schools could therefore decide how best to meet the needs of the incoming students.

A secondary science teacher entering the profession today will find a very different situation. One of the key differences is in the existence of a National Curriculum (NC) specifying the science to be taught to different age groups. The QCA determines the NC content. There is much less variety in 14–16 examination syllabuses now as any GCSE course must match the NC requirements. As science is now a compulsory core curriculum subject at primary level, students entering secondary school should have experienced, although not necessarily learned a wide range of science topics. The NC sets out topics to be taught during each of the four key stages. The curriculum is more than just a list of topics, as the PoSs detail the aspects of a topic to be covered, setting out what students should know and be able to do at the ends of each key stage.

This change in policy, from allowing schools to select teaching material to imposing a mandatory curriculum, is a serious change that all teachers should consider as a professional issue. Many entering teaching now have been taught and examined through the present system, so may assume that this is the way things are and should be without questioning the situation. Our view is that this is dangerous and that science teachers should be critical about the NC for science.

We do not think that the NC is 'wrong' or 'bad' per se; rather, the policy of having an NC needs to be explored and questioned by teachers, and the content subjected to continuing dialogue within the profession.

Who should decide what a teacher teaches?

There are clearly two extreme positions here, neither of which we would wish to advocate, which can frame a discussion of this issue. At one extreme, as teachers are performing a social and civil service, and as maintained schools are funded from public funds, then perhaps teachers should be told what to teach by the elected government, who – after all – have been assigned the responsibility to govern through the democratic process. The other extreme is that teachers are the qualified and trained *professionals*, who should be in the position of deciding what is best for the individual learners in their care.

We can see merits and demerits in both of these positions. We also suspect that some readers may think the issue is academic in view of the actual position they will face in their professional lives. Our response and starting point for suggesting how teachers should respond to the NC derive from one word featuring in the previous paragraph – i.e. that teachers are *professionals*. Although this may seem a side issue, we wish to explore what this notion of professionalism should mean, as in our view this is *central to the professional identity* that new entrants to teaching should be developing.

Although in general parlance the term 'professional' may mean little more than someone earning a living from a trade (as in 'professional plumber' or 'professional footballer'), the notion of *a profession* goes beyond this. Traditionally, the 'professions' have been self-regulating. For example, medical doctors have to register with the General Medical Council before they can practise and have done so for many years. This is more than joining a club ensuring customers receive a quality service, as occurs in many trade associations. It is a legal requirement that someone claiming to be a medical practitioner must be registered before they *can* practise medicine and that they are struck off the register and therefore prevented from practising if they are found not to be meeting professional standards. The Law Society adopts similar procedures for solicitors, who are expected to hold a current practising certificate. The certificate may be withdrawn if the solicitor does not meet appropriate professional standards and s/he may not then practise. The important point here is that the regulatory bodies come from the professions them-selves. The GMC and Law Society are professional organisations setting their own regulatory standards and acting as gatekeepers to their own register of members. If teaching is a genuine profession then it should be self-regulating in a similar way.

We have recently seen some moves in this direction. There is now a General Teaching Council for England (GTCE), with which teachers in maintained schools must be registered, and which has the power to suspend members and so debar them from working in the maintained sector (see Box 6.1). The GTCE is still relatively new, so time will be required for the organisation to establish itself as having the necessary 'teeth' to act as a true professional regulator. For the moment, many key aspects of the work of teachers are still determined by government, with – at best – consultation with the teaching unions. Our position is that if teaching is to be seen as a true profession, then *teachers, as a body,*

BOX 6.1 THE GENERAL TEACHING COUNCIL FOR ENGLAND

The General Teaching Council for England, as the professional body for teaching, provides an opportunity for teachers to shape the development of professional practice and policy, and to maintain and set professional standards.

The GTC has two main functions: to advise government on education policy, reflecting the professional views of teachers and by regulating the profession to guarantee and maintain the high professional standards that already exist.

The GTC maintains a register of qualified teachers and has the power to discipline teachers for unacceptable professional conduct or serious professional incompetence. The GTC has also issued a Professional Code of Values and Practice for Teachers describing the standards of professional conduct and practice expected of registered teachers.

In fulfilling its functions, the GTC contributes to improving standards of teaching and learning and professional standards in the interests of the public.

(Text edited from the GTC website: www.gtce.org.uk,
accessed 6 April 2004)

should be at the centre of those decisions about education which fall within our professional competence.

This does not mean we think there is no role for government in regulating teaching or education. Nevertheless, there are significant policy issues here that impinge upon teachers' working lives. Control that used to lie with LEAs has been centralised towards the DfES and other agencies such as Ofsted, the TTA and the QCA.

Neither do we think that teachers should be the *sole* managers of schools. We support the notion of school and college governance by bodies reflecting the local communities, although the balance of representation may be questioned. There are many aspects of providing an education service that should not involve teachers, such as raising taxes to pay for the service, or managing school estates. There are also key aspects of educational policy that we argue should fall within teachers' professional competence. The responsibility for making education decisions should be distributed – with levels of decision making from the profession as a body, down to the individual teacher. Between these extremes, there will be decisions that can be made by local confederations of schools, individual schools and colleges and school faculties or departments. We suggest that subject associations, such as the Association for Science Education (ASE – see Box 2, p. 15) could also be influential.

Before the statutory National Curriculum was introduced, teachers were free to decide schools' timetable structures. In most schools, students chose which subjects to study between the ages of 14 and 16. School policy would decide the balance of choices, such as whether someone could select biology, chemistry and physics or to study no science at all. In practice, most students continued with some science, but commonly for most girls this would be biology and for boys physics, closing off career options in both cases. The NC at least ensures that all students in maintained schools study 'science' to age 16, so, for example, girls cannot 'drop' physics (Taber, 1991). However, this 'advance' removes the flexibility of allowing teachers to select science content suitable for their students. In particular, schools in the 1980s offered CSE 'mode 3' courses, in which teachers could be involved in designing the syllabus. Under the NC the content is specified. Although there is a 'single science' option at KS4, allowing some students to reduce their science study to 10 per cent, the same content is deemed appropriate for all students throughout the country regardless of local circumstances or students' interests. Under the NC, science teachers who believed they were providing imaginative teaching to meet the needs of their students had, instead, to work from a detailed specification of what to teach.

This has been a major professional concern for some science teachers, who have found the NC to 'straitjacket' their teaching. The QCA claims that the curriculum time available provides opportunities for teachers to bring in other materials and ideas to meet the interests of their students, although this is not a judgement many science teachers would accept. Even if the NC could be taught in a fraction of the available time, this would not address the question of whether *the content* of the curriculum is suitable material for all students. The KS4 NC for science is in the process of being reviewed, but only time will tell whether professional science teachers will judge the new curriculum an improvement in terms of matching the needs of many students.

We raise this issue here because we think this is an important concern. In what follows we will discuss some of the decisions that a teacher must make about *what* to teach, as well as how to teach. We think it vital to acknowledge there is a genuine professional issue over the extent to which such professional decision making is constrained by external stipulations.

DEVELOPING SCHEMES OF WORK

Making initial plans

A curriculum or examination specification is *not* a plan for teaching. These documents describe the content or skills to be included in a course, but do not specify how or in what sequence material should be taught. They provide an agenda or checklist of what students are to be taught. In the case of the NC – an official policy document – the 'programmes of study' (PoS) are mandatory. Strictly speaking, an examination syllabus does not set out what a teacher should teach, but rather specifies what students could be examined upon at the end of the course of study.

The process of planning to teach starts from the curriculum document and moves through stages. A new entrant to the profession, such as a trainee teacher, is normally inducted into the process by being asked to first plan lesson episodes, later lessons and sequences of lessons, while being advised (and censored) by a mentor – an experienced qualified teacher. This is a sensible approach in terms of the new entrant's own learning to teach, but this short-circuits the way planning tends to move from the long term, through the medium term, to the short term. In other words, faced with a course to teach, and the associated curriculum document, it is *not* sensible to start by planning individual classroom activities or even whole lessons.

Sensible planning starts by considering the time available, and the total amount of material to be taught, and then developing an overall outline plan for the course.

For example, consider a two-year GCSE course leading to examination at age 16. A decision must be made about how the syllabus material should be organised across the duration of the course. This is not a trivial matter. A science department will make decisions about such matters as:

- the number and length of modules, units or topics of work; and then,
- how the material should be divided into modules, units or topics that will have some coherence; and then,
- the order in which the topics should be learnt.

In practice, there are likely to be practical constraints relating to time-tabling, staffing and room use. These can be major issues within a school, sometimes creating major problems for the Head of Science. These are normally beyond the experience of an inductee, but it is important to be aware of them as they may constrain what is practically possible. To consider one example, teachers may wish to teach topics in a particular order (discussed further on p. 128). However, if several classes have science lessons at the same time, there may be limitations on apparatus, so topics have to rotate through the groups. This means that some classes do not meet topics in the preferred order.

What is included in a scheme of work (SoW)?

Schemes of work vary in their level of detail. They should not be too prescriptive, as this undermines the imperative to plan lessons to meet the needs of a particular class. An inductee may, however, appreciate more, rather than less, detail – as long as this is largely seen as *guidance* and not a restriction on professional decision making.

In England the QCA publishes 'model' schemes of work on its web-site (www.qca.org.uk/, accessed 1 August 2004, discussed on pp. 69–70).

Although these are not mandatory, they indicate what the government's agencies think should be included in a scheme of work. This is helpful when seen from the perspective of school inspectors, who will expect to see thorough and careful planning. Many schools use these schemes of work as the starting point for developing their own school schemes, although some detail in the QCA schemes suits individual lesson planning rather than SoWs which should be limited to overviews of what material is to be taught in which lessons.

The QCA schemes are useful in reminding us of key aspects to be kept in mind when planning teaching, for example:

- continuity and progression;
- differentiation;
- assessment.

Each of these themes should influence a teacher's planning decisions, so it is worth saying a little about each at this point.

THEMES IN PLANNING TO TEACH: CONTINUITY, MEANINGFUL LEARNING AND PROGRESSION

The learner's experience of continuity in learning

As a teacher you have a privileged viewpoint on what happens in your classroom – you have planned the lesson, and (hopefully) understand in advance not only the material to be taught but also the rationale for the lesson activities. The learner's perspective is very different. Most children enjoy school, but for many this is probably because of the social aspect: many of the students you teach do not primarily come to school because they particularly want to be in your lessons. Of course, some students will love your subject, and many more will be won-over by your respect and interest in them, your fascination and enthusiasm for your subject and your hard work in making lessons interesting and fun. However, for many secondary students school is an on-going adolescent drama, broken up by those frequent, but somewhat discrete, interruptions that we call lessons.

Students coming to your lesson usually have a great deal on their minds apart from learning science; have had plenty of time since your last lesson to forget what it was about; and have probably experienced (whether enjoyed or endured) a number of very different lessons in other subjects since you last saw them. As a teacher, especially a new teacher, there is a great temptation to enter the classroom seeing the start of this lesson as the natural continuation of where you left off last lesson. With most classes, this mind-set will not be shared by the majority of the students. You need to plan to provide the sense of continuity that is important if your students' science experience is not to be a series of disjointed lessons. You need to think carefully about how each lesson follows on from what has gone before, and then think about how you can make the connection clear to the students.

The implication here is that sequences of lessons have to be planned to provide an effective thread that can provide learners with the sense of continuity they need to make sense of a subject. One way of thinking about this is to develop a narrative that flows through the sequence of lessons on a topic (cf. comments about storylines in the Salters' courses, see Chapter 4). Although such a narrative may seem to *over*simplify the topic you are teaching, this may provide *a framework* around which students can organise their learning (Mortimer and Scott, 2003). Once the students have acquired this framework, you can then develop their understanding of the complexities and subtleties of the topic.

Meaningful learning

A key distinction in learning is between what has been described as 'rote' and 'meaningful' learning (Ausubel, 2000). Rote learning is 'learning-by-heart', and meaningful learning could be characterised as 'learning-for-understanding' (see Box 6.2).

The meaning of any concept, such as photosynthesis, oxidation or refraction, depends on how we link that concept mentally to others that we consider related. We can think of the mental representation of concepts as a large net or web, with each node being a concept and the connecting threads as the relationships between them. Each learner's meaning for a concept can be considered the sum total of all the links with other concepts, including the indirect ones. Of course, this image does not reflect how some links may be stronger than others, but it does show how concepts are interdependent and how each individual's meaning for any concept is to some extent unique. For example, a year 11 student who knows that metals tend to be sonorous, malleable and ductile has a different concept of metal from a year 9 student who knows

BOX 6.2 MEANINGFUL LEARNING

The psychologist David Ausubel is known for stating: 'Find out what a learner already knows and teach accordingly.' Ausubel introduced the notion that learning needed to be meaningful and that this depended on the learner's cognitive structure and the nature of the material to be learned (Ausubel, 1961). Ausubel and Robinson (1971 [1969]) suggest three conditions for meaningful learning to occur:

- The material must be relatable to some hypothetical cognitive structure in a non-arbitrary and substantive fashion.
- The learner must possess relevant ideas to which to relate the [new] material.
- The learner must possess the intent to relate these ideas to cognitive structure in a non-arbitrary and substantive fashion (p. 53).

If new material presented to a learner cannot be related to existing knowledge, then it cannot be learned in a meaningful way, so must be learned by rote if it is to be learned at all. Second, if meaningful learning implies some form of integration of assimilation with existing knowledge, then if the existing knowledge contradicts that of the authority (teacher, textbook) the new material will not be understood in the way intended.

only that they conduct electricity, while a year 13 student may know about Young's modulus and why metals show elastic behaviour.

There are times when we all need to learn by rote, such as remembering PIN codes, but we contend that education should largely be about *learning-for-understanding*, that is, meaningful learning.

New learning is meaningful when it is related to existing learning and the more it is linked to existing knowledge, the more meaningful it becomes. One description of teaching might be 'making the unfamiliar familiar' (Taber, 2002b) and part of the teacher's role is to make learning meaningful for students. The teacher has to show students how the abstract concepts they learn in the curriculum relate to things that are familiar and, where possible, important in their lives. Making students see the relevance and significance of scientific ideas makes the science more interesting, and helps motivate them to learn.

In Chapter 3 we talked about the rationale for teaching science. We do not want students to learn science as unrelated facts, but as a set of ideas that helps us understand, relate to, and sometimes control, the world in which we all live. We want students to acquire scientific understanding that they can apply widely, not just to examination questions at the end of their courses. The more material is linked in learning, the easier it is for the ideas to be accessed from memory later (Taber, 2003d).

These considerations suggest that the teacher should seek to make as many relevant links as possible in their teaching. This does not just mean links with the prerequisite knowledge underpinning a new concept, but with other aspects of the topic, the subject and the curriculum. Experienced teachers often introduce suitable links spontaneously in their teaching, but for the new teacher explicit inclusion in the planning process is the best way to start.

Progression in learning

Progression is an essential aspect of students' learning. This term implies that students do not just learn more of the same but, rather, that they learn at higher levels. As well as developing more integrated knowledge structures about science, students are expected to gradually deal with more depth, detail, complexity and subtlety and to demonstrate higher level skills.

A well-known system for classifying the relative difficulty of different cognitive skills is known as 'Bloom's taxonomy' (see Box 6.3). This suggests that whereas recall of knowledge and simple comprehension tasks are fairly low-level skills, asking students to analyse, synthesise and criticise sets much greater demands. Some perspectives on how children's cognitive abilities develop suggest that many secondary students will find higher level tasks at least difficult, if not impossible (Shayer and Adey, 1981). However, this is probably a simplistic view and is no excuse not to challenge and extend students. The NC for science was initially set up with ten hierarchical levels each with descriptors to illustrate the thinking students should demonstrate to be 'awarded' the different levels. The original version was found to be overly optimistic, as the top levels were found to be unrealistic for the vast majority of secondary age students.

The NC levels are a genuine attempt to ensure that progression through school means more than just learning more-of-the-same. Some formal assessments are designed to assign students to particular levels and secondary schools are provided with information about the level

BOX 6.3 BLOOM'S TAXONOMY

Bloom and colleagues produced taxonomies of education objectives in the cognitive and affective domains. Reference to Bloom's taxonomy usually signifies the taxonomy of educational objectives in the cognitive domain (Bloom, 1964). Bloom proposed six levels of demand associated with tasks that learners might be set. These are: (1) Knowledge; (2) Comprehension; (3) Application; (4) Analysis; (5) Synthesis; (6) Evaluation. Numbers 2–5 were collectively considered 'Intellectual abilities and skills'. Although this original version of the taxonomy is well known, a revised version has been developed as a 'taxonomy for learning, teaching and assessing' (Anderson and Krathwohl, 2001), separating out the types of cognitive process applied to knowledge and the type of knowledge acted upon (Krathwohl, 2002). In the revised taxonomy, the main classes of cognitive processes are:

1. Remembering: recognising, recalling
2. Understanding: interpreting, exemplifying, classifying, summarising, inferring, comparing, explaining
3. Applying: executing, implementing
4. Analysing: differentiating, organising, attributing
5. Evaluating: checking, critiquing
6. Creating: generating, planning, producing.

The knowledge dimension of the revised taxonomy has four main categories:

A. Factual knowledge
B. Conceptual knowledge
C. Procedural knowledge
D. Metacognitive knowledge.

Any learning or assessment task can be classified in a cell of the grid made up by tabulating the cognitive processes, for example, as columns, against the knowledge dimension as rows.

assigned to new students arriving from primary schools. This information is meant to help inform secondary teachers so that they can pitch lessons at an appropriate level. In practice the system does not always work well (Herrington and Doyle, 1997; Hargreaves and Galton, 2002), so many secondary schools use their own tests to assign levels rather

Table 6.1 *Extracts from the Sc1 level descriptions*

NC level	Extracts from level descriptions for Sc1, KS3
3	. . . They make relevant observations and measure quantities such as length or mass using a range of simple equipment . . .
4	. . . They select suitable equipment and make a series of observations and measurements that are adequate for the task . . .
5	. . . They select apparatus for a range of tasks and plan to use it effectively. They make a series of observations, comparisons or measurements with precision appropriate to the task . . .
6	. . . They make enough measurements, comparisons and observations for the task. They measure a variety of quantities with precision, using instruments with fine scale divisions . . .
7	. . . They make systematic observations and measurements with precision, using a wide range of apparatus . . .

crudely (see the discussion of criterion referenced assessment in Chapter 10). Nevertheless, the NC level descriptions provide criteria for judging whether learners are progressing in their science learning. For example, Table 6.1 presents brief extracts from the Sc1 descriptions for levels 3–7, the range appropriate for most KS3 students. The full level descriptions cover the whole PoS for science, but Table 6.1 provides brief extracts showing comparable points. Consider a student who was assessed as barely able to '*make relevant observations and measure quantities, such as length or mass, using a range of simple equipment*' (level 3) on starting secondary school. This student would be judged as making substantial progress in this aspect of Sc1 if three years later s/he were judged as being able to '*make systematic observations and measurements with precision, using a wide range of apparatus*' (level 7).

MEDIUM-TERM PLANNING

Planning for assessment

Assessment is an essential aspect of teaching and learning, because without it neither teacher nor learner has a clear indication of the learning that might or might not be occurring. To many inductees 'assessment' may be associated with formal tests and examinations. We will see in Chapter 10 that this is only one aspect of assessment. Assessment needs to be considered in all stages of the planning process and it is worth pointing out here that SoWs should indicate when formal assessment will occur. External assessment dates are set by QCA or examination boards well in advance, so schools can plan teaching course material around them.

Planning the topic

The SoW will outline the order in which topics are taught. Within each topic, the SoW will specify an order for teaching the topic contents. Sequencing of material is much more critical in some school subjects than others. In some subjects there may be a range of topics of similar complexity through which a range of central concepts and key skills are taught and learnt. For these subjects the order of topics is quite flexible, although once decided then teachers must ensure there is progression in learning as students move from topic to topic.

For example, one of the authors worked with a colleague who taught the humanities subjects history, and government and politics at A level. At the time A levels were not divided into AS and A2 levels and were usually assessed by terminal written examinations. In order to give viable group sizes this teacher taught her subjects on a rolling programme, teaching a different set of topics in alternate years. This meant that in any group there would be first and second year students, experiencing a completely different sequence of topics. The teacher was able to do this because the key skills and concepts of her subjects could be taught through any of the topics. She expected second-year students to demonstrate more sophisticated understanding than their first-year peers. The nature of the subject matter made this possible; the author never attempted the same approach in his A level physics and chemistry classes. In Chapter 7 we discuss sequencing of science topics.

Principles that inform lesson planning

So far, this chapter has looked at planning in the long and medium term, largely talking about planning at the course and topic levels. The Head of Department, or the department teachers as a group, undertake most of the planning at this level. Individual teachers, however, are normally responsible for planning the actual lessons.

Lesson planning will use the SoW (and where appropriate either the NC or examination specifications) to indicate the topics to be explored in a particular *lesson*. We all probably have some strong memories of what lessons were like when we were at school, but it is nevertheless common practice to expect applicants for places on PGCE courses to have spent some time observing school science lessons before being interviewed for a place in teacher training. As well as providing a context for answering questions at interview, this also reminds the applicant what lessons are like seen through the eyes of an adult, rather than as a student. Even having done so, some inductees find it hard to appreciate what happens in a typical lesson. Rather, probably as a result of many hours of teaching received in higher education, there is a tendency to think that preparing for a lesson means mastering the subject matter, and teaching the lesson means communicating this to students. The traditional model of university teaching can influence new entrants' perceptions of the whole teaching process. We offer a caricature of this in Table 6.2, contrasting 'naive' and 'effective' views of teaching:

Although some university lectures experienced by the authors appear to have been designed along these lines, most lecturers probably put much more effort into their preparation than students expect. Also, it is useful to note that:

- university lecturers are traditionally employed for their research potential and until recently had little if any training as teachers;
- universities have traditionally always been judged on the quality of research, rather than teaching quality, although this has now also changed;
- lectures are considered an effective means of communicating information to relatively large numbers of students, all of whom have been selected for the courses on the basis of aptitude and proven ability to study;

Table 6.2 Two perspectives on teaching

Activity	Naive view (lecturing)	Effective view (school teaching)
Preparing for teaching	Mastering the content	Transforming the content
Teaching	Communicating the content	Facilitating learning
Assessment	Testing the students	Informing teaching

- lectures are just one aspect of university teaching; others include supervisions, tutorials and laboratory classes.

Also, many universities are moving away from the traditional lecturing model to include more interactive lectures, workshops, problem solving and project work. Conferences such as the *Variety* meeting explore innovations in university chemistry teaching (www.dbweb.liv.av.uk/ltsnpsc/variety.asp accessed 28 April 2004).

These points illustrate that a university lecture cannot be adopted as a suitable model for a science lesson in a secondary school. Although few inductees are likely to suggest lecturing as an appropriate teaching approach with 11-year-olds, the many hours of exposure to university lectures seems to provide trainees with a template for what teaching should be, that is, a lecture. Commonly, inductees talk too much in their lessons, leaving time for the students to do too little. Students get fidgety after only 15 or 20 minutes of teacher talk, as this involves them having to listen. Inductees are also often shocked at how little content is covered in a one-hour lesson even by experienced teachers. Those who avoid such errors with younger students tend to revert to lecturing when working with year 12 and year 13 groups.

In practice, very few effective science lessons are structured so as to be a straightforward communication of material from teacher to students. If teaching were that simple we could surely video some excellent teachers explaining science and mass-produce exemplar lessons on tape or CD-ROM at a tiny fraction of the cost of training and employing teachers.

Teaching *is* about bringing about learning, which is something more than passing on information. Learning is a very complex set of processes. We now know enough about how learning works to realise that each individual student has to effectively build up their own understanding of material internally as a *personal* act of knowledge construction (e.g. Pope, 1982). This personal act is usually only effective if it is socially mediated, for example, by the teacher or through interactions with peers (Solomon, 1987), but is a unique event. The individual nature of learning depends upon the learners' cognitive apparatus – that is, how individual perceptual, thinking and memory systems work and their learning styles, by which we mean their preferred ways of learning material. Moreover, we know that the biggest determinant of what a person will learn from

Table 6.3 *Key points about learning*

Learners vary considerably along many dimensions. An effective learning opportunity for one student or group may be poor use of time for another.

Meaningful learning takes place when the new information or activity can be understood in terms of prior learning or experience.

Students can only mentally juggle a small number of factors at any one time: in general the complexity of material, as seen from the students' resolution, is much greater than it appears to the teacher.

Lessons that students enjoy are more likely to be remembered and to lead to effective teaching.

Students bring their own alternative conceptions of many curriculum topics to their lessons and these ideas may be more tenacious than the material met in class.

Developing new knowledge requires active processing of information – through talk, composing text, forming and exploring questions and application in familiar and novel contexts.

Students have different learning and thinking styles and in any class many of the students' preferred ways of learning and thinking will not match the teacher's.

Information presented in several modes (spoken, written, diagrammatic, etc.) is more likely to be assimilated than information in a single mode.

Students cannot focus on learning if they are cold, hot, tired, hungry, thirsty or scared, etc.

Learning within the brain is an on-going process and some significant conceptual change may take months or longer: major concepts will not be learnt by one-shot teaching.

Directed activities related to text (DART: sequencing, completing, relating, translating) are more likely to lead to learning than copying of notes. Students take their brains, but not usually their notes into tests and examinations and into adult life after school.

Activities that are too easy are boring; activities that are too difficult are frustrating and damage confidence and self-image. Activities that are challenging but structured to lead to success and opportunities for praise will bring about learning and improved attitudes to school work.

Talk between peers can often provide an opportunity to develop thinking about a topic at a level and pace that suits the students.

Students will tend to interpret new information in terms of any existing alternative conceptions and this can happen without the teacher being aware unless probing questioning is used.

Lessons that are planned without careful content analysis will often assume too much prior knowledge proving incomprehensible to students and frustrating to both teacher and students.

Even when new ideas are introduced in the spiral curriculum, students will find the material boring if the context and content is too similar to that used in earlier key stages.

Eliciting prior knowledge prevents optimistic and costly assumptions about what *is* already known and sometimes prevents boring repetition due to pessimistic assumptions about what is *not* already known.

a lesson is what they already know, as this will determine what and how they understand and where they have appropriate foundations to build new knowledge (Taber, 2003e and Table 6.3). Given that every student in the class will have different background knowledge, each student will learn something different in a lesson. Research indicates that many students hold significant alternative understandings of key science concepts (or 'alternative conceptions', see pp. 102–7). It is likely that the extent to which new learning matches the teacher's intention will vary considerably from student to student (Taber, 2001a).

There is considerable research on how learning occurs. We strongly recommend that science teachers familiarise themselves with this material. We do not, in this broad-based book, have space to do justice to the important issues surrounding the factors influencing learning. However, we cannot over-emphasise the importance of understanding the learning process when planning effective teaching. Table 6.3 gives a list of key points on learning which teachers planning lessons may like to keep in mind.

THINKING ABOUT PRACTICE

- Who should decide the Curriculum? Who do you think should decide what students should learn? There are many possibilities: should it be national government, the local education authority, the school's governing body, the senior management team, a representative school curriculum group, the Head of Department, the department as a group or the individual teacher? Perhaps you think the students themselves should sometimes have a say in what they are to learn. Is there a case for the decision making to be distributed across a number of these levels?

- If a government agency has analysed and sequenced the NC at KS3, and prepared schemes of work with detailed units that are considered to act as exemplars, then why might an individual school department decide to design and develop their own schemes?

- What needs to be *taken into account* when planning teaching? In particular, are there different areas ('domains') of knowledge that a teacher needs to plan effective teaching?

- Might there be a good rationale for the teacher not 'covering' all the material specified in the examination 'syllabus' with a class?
- In what way(s) is a teacher better prepared to make sense of the material s/he is teaching than the students who will be experiencing the lesson?
- Choose three points from Table 6.3 that you think are the most significant in planning a science lesson. Discuss your choices with colleagues. Do your choices reflect your own personal learning styles?

Chapter 7

Planning to teach the science

The medium is the message.
(Marshall McLuhan)

INTRODUCTION: HOW DOES THE NATURE OF THE SUBJECT IMPACT UPON THE PLANNING PROCESS?

The previous chapter was about planning to teach the curriculum, and the principles would be largely applicable to any subject area. The present chapter builds upon the last, by considering how we apply these ideas in the particular context of science teaching. To some extent that means that this chapter is generally focused at a 'finer grain size' (i.e. more about planning individual lessons), but the main theme of this chapter draws upon the material in Part I of the book. Here, we are concerned with asking *how the nature of science as a school subject influences the way we go about preparing to teach.*

The key points listed in Table 6.3 at the end of the previous chapter are presented as 'tips for teachers', but they are the distillation of a great deal of research into how learning occurs. During a course of teacher training, an inductee can expect to be introduced to some of the background behind these points – the research on which they are based and the ways teachers operate with these ideas in the classroom. Some of these points are also reflected in the present chapter where we consider the process of planning to teach the science.

THE STRUCTURE OF SCIENTIFIC KNOWLEDGE

Science topics are interlinked inextricably

First, we need to understand *the structure of the material* to be taught. Science as a subject has a highly non-arbitrary structure. Although science as represented in journals and books is socially constructed (see Chapter 2), nonetheless it is highly constrained by nature. The way we choose to define 'elements' and 'compounds' bears imprints of the historical process through which these concepts have developed, so the concepts do reflect human nature. However, there is little doubt that the basic distinction reflects a real and important regularity found in nature. There are undoubtedly things in science that could have been done differently and which would work more-or-less as well, but that does not mean 'anything goes'.

SCIENCE builds models of the world. The models are not nature itself but representations of aspects of nature. Moreover, they are imperfect human representations of particular abstractions that seem significant to the human mind. Yet SCIENCE is successful because it has developed effective ways of building, developing and censoring the models it uses. Models are retained for as long as they have utility – for example, in helping us to explain and predict, or suggest fruitful ways to move research forward. As Chapter 2 indicates, models that are no longer useful in SCIENCE do sometimes move on to a new career in the science curriculum.

One of the ways that scientific ideas are judged is in terms of consistency. Theories are expected to be internally coherent, so they do not contradict themselves. SCIENCE also usually shows limited tolerance of ideas that seem inconsistent with other areas of SCIENCE that are well established. The nature of SCIENCE, then, is that it comprises entities (theories, models, explanations) constrained by internal logical requirements and often highly interconnected.

We need to bear this in mind when we teach science. Topics from the science curriculum – respiration, polymers, gravity – do not stand alone. We can discuss any topic with others who also have science knowledge and appear to stay within the topic. Yet such talk would almost certainly draw upon other ideas needed to make sense of what we say, that we have not openly acknowledged and explored. For example, explaining respiration would be hard without using the concept of energy, polymers without molecules, or gravity without assuming a mutual understanding of 'force'.

We are sure that specialists in most curriculum areas would make similar claims: that there are basic ideas used to structure teaching and learning. We would argue, however, that there is a great difference in degree. *In science, the linking and mutual support between concepts is ubiquitous and fundamental to the subject.* A new 'theory' of radioactivity that did not seem to relate to established thinking about energy and atomic structure would not be considered very seriously, no matter how well it seemed to fit the facts in its own range of application. Only when existing theories make little progress in explaining known anomalies are extreme solutions (such as quantum theory) given a good hearing.

The lingua franca of science?

We suspect that most science graduates would readily agree with the perspective outlined above, not finding our description of the structure of scientific knowledge very remarkable. This is just part of our collective and largely tacit appreciation of the way science is. When we talk to other science experts about science topics we can take it for granted that they understand the concepts and principles that underpin the topic being discussed and this is usually a safe assumption.

The situation is different when talking to students. By definition, they are not yet experts in science, so we cannot assume automatically that they will either know the underpinning knowledge or, indeed, actually recognise it as relevant when they do. This is clear to most inductees into science teaching.

However, acknowledging that one has to be careful in the assumptions we make about students' existing knowledge is one thing, but avoiding making errors is quite another. We suggest that some ideas used in science have become so ingrained by the time one attains a science degree that errors are not easy to avoid.

For example, the idea that all materials are comprised of myriad tiny particles is fundamental to most areas of modern SCIENCE. Yet *this* is an area that students find difficult to make sense of (e.g. Taber, 2001b). An inductee may give an explanation based implicitly upon this principle (or others such as the conservation of energy, or Newton's third law, etc.) without even realising this is the case. Trainee teachers often get into difficulties because they have not realised that their careful explanation only makes sense in the context of some basic scientific idea that students either do not have, or at least are unlikely to apply spontaneously. Changing the direction of an explanation in the light of this realisation, or introducing whole areas of underpinning knowledge can complicate and lengthen the intended presentation and upset timings in lesson plans.

Avoiding these difficulties is very hard because some of our science knowledge has become so well integrated into *our* thinking that *we* spontaneously access and apply it with minimal effort. This is linked to the way memory is consolidated over time, which allows quite complex information to be accessed and manipulated very easily once it has become very familiar. It is as if some of these ideas and principles are no longer treated as conceptual knowledge, but become part of our everyday vocabulary. Science graduates readily talk in a specialised language that makes little sense to most of the rest of the world. Being a science teacher in a class of students is like teaching a group who do not share your first language: everything you want to say has to be carefully translated first (Taber, 2002b).

Content analysis: the translation process

Clearly, translating our science to the language available to, say, typical 14-year-olds is very difficult to do in real time. Part of preparing to teach science, then, is to plan this translation process in advance. Any science teacher faced with teaching a topic, at any level, for the first time, is well advised to start the process of 'translating' a scheme of work into lesson plans by undertaking a thorough 'content analysis'.

The process of content analysis involves:

- breaking each idea down to see how it is explained in terms of other more fundamental ideas;
- identifying all the links between aspects of the topic.

We might see these two aspects of the process as the analytical and synthetic parts of the overall process (Taber, 2002b). Both are important to effective teaching, but we suggest the analytical aspect is essential if we are to be confident that our students are likely to understand what we are talking about.

137

The analytical aspect: identifying prerequisite knowledge

The analytical process is about identifying the concepts that we will use (explicitly or otherwise) to build up new knowledge. Consider as an example the idea that in chemical terms a metal is an element that tends to form cations (perhaps linked to the section on the periodic table under 'patterns of behaviour' at KS4 – see Box 7.1). This information would only make sense to a student who has already acquired a concept of chemical element, and understood something about atomic structure. It would clearly not make sense to try to teach *this* meaning for the term metal to any student who did not already have prior learning about elements and atoms in place.

BOX 7.1 THE PERIODIC TABLE

Pupils should be taught:

- the connection between the arrangement of outer electrons and the position of an element in the periodic table;
- that elements in the same group of the periodic table have similar properties.

(From the Sc3 PoS at KS4, DfEE/QCA, 1999)

This is just one example, but the principle can be repeated for all aspects of science. Perhaps similar points can be made about other subjects, but we think this is much less of an issue in most school subjects. Sequencing of ideas is essential to making sense in science because scientific knowledge is highly structured, often in hierarchical ways. Sitting with an SoW and identifying the prerequisite knowledge is a sensible investment as part of preparing to teach science. When a teacher is considering how to transform knowledge through teaching, the complexity of the components of knowledge should be identified in the analysis.

People have been shown to have very limited 'working memory' – they can only deal with limited data at any one time (Miller, 1968). As material becomes familiar the brain's cognitive apparatus is able to treat increasingly complex information as if it were more primitive. In other words, the memory capacity operates in terms of the *perceived* difficulty of the material: the information content in something like '$CH_3.CH_2.CH_2.COOH$'

varies according to the individual's familiarity with the material. Reproducing $CH_3.CH_2.CH_2.COOH$ is trivial for a chemist: identifying $CH_3.CH_2.$ $CH_2.COOH$ as butanoic acid allows the information to be almost effortlessly processed and reproduced. The same string of symbols would overstretch the memory capacity of most year 7 students who would have to learn it by rote.

The ability of our brains to 'chunk' information about familiar items is probably essential to our mental lives, but causes a problem for teachers. The science inductee has to re-learn the complexity of familiar information from the perspective of the students (Taber, 2002b). A class can very easily become confused because a 'simple' explanation to the teacher is much less simple when considered by students who are less familiar with the topic.

The synthetic aspect: concept mapping

As well as looking for prerequisite knowledge, teachers must also consider how aspects of a topic fit together. Sometimes we build knowledge by sequencing material, so that we use ideas we teach early on as the basis for developing new understanding later. This early information then acts as prerequisite knowledge for the later material. However, even when this is not strictly the case, exploring how the parts of the topic link together can be useful. Some links may be needed in a later topic, or when studying the same topic at a higher level later in the spiral curriculum, but are unlikely to be obvious to the students. Again, we have the problem that our familiarity with the science can act as a barrier to appreciating the way the topic appears to the learner. Even when the links are not essential, drawing upon them may make good sense. We suggest three functions for links:

- narrative;
- reinforcing;
- anchoring.

There are almost certainly others.

As students do not enter the topic having the overview that the teacher brings, we have argued that it is important to give them a sense that what they are learning can have some kind of coherence and pattern. We suggested that in teaching a topic we might almost be looking to provide some form of narrative that engages the students.

Links within, and between topics, also help to support the learning process. Reinforcement is important because memory is a selective process. Material from short-term memory stores (believed to be of the form of electrical oscillations) may be transferred into some more permanent form (i.e. in circuits of nerve cells modified by changes in synapse strengths). Even when such permanent traces are laid down, they are likely to be weak (i.e. have a relatively high threshold for activation) unless they are re-activated regularly soon after initially being established (Anderson, 1995). This is the basis for suggestions in self-help revision books that learning should be reviewed after one hour, one day, one week and one month, etc.

Taking opportunities to review recent learning where it can be made relevant will reactivate these memory circuits, which then strengthens the connections and makes them more likely to be accessed in the future (Sousa, 2001). Linking with new material also helps as it provides alternative pathways to the stored material, which makes recall more likely.

This linkage also supports learning in the opposite direction – from established to new learning. Learning that is meaningful (see Box 6.2, p. 124) is more readily accessed and is more likely to be applied in the intended contexts. Scientific facts that are learned by rote are like islands of knowledge in uncharted waters: out there somewhere but only likely to be found by an act of luck. Material that is understood in relation to existing knowledge, i.e. material that makes sense to the learner, is more likely to be stored in permanent memory, and much more likely to be accessible in scientifically relevant contexts. By linking new material to existing learning it is given a context and is anchored into the relatively stable aspects of the students' existing knowledge (Ellis and Hunt, 1989).

There is no single way of undertaking the content analysis to identify relevant links, but concept mapping is to be recommended (Al-Kunifed and Wandersee, 1990; Novak, 1990). Concept maps are diagrammatic representations of areas of knowledge that allow links to be readily drawn between concepts. Concept maps are used in several different ways: purely for the teacher's personal preparation; to provide a route map of a new topic for students; or as a summary at the end of a topic. They can also be used as an alternative to linear notes for learners who find them easier to make sense of, or partially completed in the form of DART (directed activities relating to text) activities. Concept mapping can also be used as a means of eliciting students' prior ideas (individually, or as a mind-storm activity), or as a study/revision technique for students – or a way to encourage more creative thinking about topics (Taber, 2002b).

140

PLANNING FOR SCIENCE TEACHING FROM STUDENTS' PERSPECTIVES

Researching known aspects of students' thinking about science topics

Another aspect of preparing to teach, which – if not unique – is certainly of particular importance in science is finding out about research into learners' ideas about the topic. There is pedagogic knowledge in all teaching subjects that considers learning difficulties in different aspects of the subject – if only at the level of why students find certain aspects of a subject difficult.

In *science* education, however, there is also an enormous literature on learners' ideas (and especially their alternative conceptions) in science. For example, the book *Making Sense of Secondary Science* (Driver *et al.*, 1994) reviews the major findings of this research in the science topics in the secondary science curriculum.

There is no doubt that the literature in this area lacks coherence. Different philosophical stances to research and knowledge, different methodologies, and different ways of describing findings, all complicate the picture.

What does seem clear is that:

- students arrive in science lessons with alternative understandings of many phenomena that they will study in science;
- some of these alternative understandings are tenacious;
- some 'alternative conceptions' are commonly held;
- students may retain their original ways of thinking despite teaching;
- eliciting and responding to students' initial ideas can lead to success in getting them to change their minds.

It is important to acknowledge some of the other conclusions one can draw from the vast literature available:

- sometimes students show limited commitment to their initial ideas and no special attention would be needed to persuade them to switch to the desired way of thinking;
- in many topics, there are reports of learners' ideas, perhaps with suggestions of how to respond, but little evidence of successful

141

strategies having been tested (there are many more studies reporting conceptions than bringing about changes, for the simple reason that research to uncover alternative conceptions is much easier to undertake).

Science teachers cannot obtain definitive, well-researched strategies for challenging students' ideas in all topics from the research literature. However, teaching informed by a knowledge of common alternative conceptions and a broad awareness of the wider literature describing, in general terms, how to respond, is likely to be more successful than teaching that pays no attention to this issue. We strongly recommend that planning for teaching science should involve exploration of research into learners' ideas, thinking about how and why these arise and how the scientific evidence available can be used to *persuade* students to accept and learn the models of the science curriculum.

Ogborn and colleagues (Ogborn *et al.*, 1996) have discussed how entities of SCIENCE have been created through scientific discourse, both through interrogation of nature and the subsequent rhetoric of the scientific literature. With students we must short-circuit this process considerably, but we still need to re-create the conceptual entities of SCIENCE for them, and with them, through persuasive science lessons.

Avoiding 'pedagogic learning impediments'?

There is a danger inherent to the process we are discussing that merits a pause for thought. Bringing together some of the ideas we have been emphasising in previous chapters, we see that in trying to re-create the entities of SCIENCE we (collectively as a profession, as well as individually) may inadvertently create and propagate something quite different. The term 'pedagogic learning impediment' has been used to describe ideas that act as a block on acquiring scientific understandings, but which themselves *derive from teaching* (Taber, 2001a).

Becoming a science teacher involves taking on a new role and professional identity. Science teachers are not involved in cutting-edge SCIENCE and neither can they be expected to have scientific knowledge in depth and breadth across the curriculum. We have said also that due to recruitment problems, many schools may not have the breadth of in-depth subject knowledge across the whole of science in the science department. Science teachers' specialist subject knowledge, then, is not that of science per se (something that would be beyond any single human being in the twenty-first century), but of the entity we call school science. School science is in effect a set of *simplifications* of aspects of the sciences, designed to be accessible to secondary students and socially constructed through the political process leading to the NC. Science teachers work from the curriculum, designing lessons to teach these models. As well as their own background knowledge they draw upon a range of resources: the departmental SoW, more experienced colleagues and ideas in various textbooks, commercial schemes and support materials.

Preparing a lesson in a topic when you have a high level of up-to-date scientific knowledge enables more accurate judgements to be made about the curriculum models and teaching approaches in terms of how well these reflect the SCIENCE. Teaching a topic in which you have less confidence in your scientific knowledge brings a tendency to seek teaching ideas, and it is here that textbooks, experienced colleagues and commercial resources can be regarded as sources of authority. We think that under these conditions there is a considerable danger of curriculum models taking on a 'life of their own'.

We talked in Chapter 2 about the idea of 'conceptual fossils' being found in the curriculum: ideas that ceased to have scientific value decades previously, but which are still taught. The literature on learners' ideas in science suggests this is not idle speculation. In Chapter 5 we discussed how many students have intuitive ideas about force and motion that are acquired before they start schooling. Many learners' ideas seem to be 'intuitive' in this way, or are likely to have been acquired from social and linguistic cues. For example, unsurprisingly, many students think that plants get most of their material from the soil – supplemented by 'plant food'. Pause here to ensure you understand what is wrong with this notion. Yet, there are some common alternative conceptions that seem unlikely to have their origins in this way.

As an example consider common ideas about why chemical reactions occur. When year 12 and year 13 students are asked why hydrogen and chlorine react they commonly claim this is because the hydrogen and chlorine atoms 'want to have full shells of electrons and so share to fill their shells' (Barker, 1994; Taber, 2002b) This is the common response, even when the question includes the reaction equation:

$$H_2 + Cl_2 \rightarrow 2HCl$$

From an objective point of view, it seems very odd that 16–18-year-olds should suggest an explanation that is clearly inconsistent with the information given in the question. That three-quarters or more of a class may present answers along these lines seems incredible. Even if the students had no idea of a valid explanation, suggesting such an invalid explanation seems highly unlikely. Unlike the 'force and motion' example, though, this can hardly be due to 'intuitive' science – without teaching, minds are hardly likely to construct a theory of chemistry that explains chemical reactions in terms of the desires of atoms.

The source of students' difficulties is most likely to be a combination of the nature of science as a source of explanations, and the lack of any suitable curriculum explanation for why reactions occur that is considered appropriate at secondary (i.e. KS3 and KS4) level. Although explanations are provided at sixth form level that are authentic in terms of the chemistry, most students already have 'acquired' the 'atoms want/need full shells' notion by the time they reach the end of compulsory schooling.

With time the 'explanatory vacuum' at secondary level has become filled with the notion of 'full shells' as an imperative. Many teachers actually teach this idea supported by textbooks that at least imply (and sometimes are quite explicit) that chemical reactions *occur* for this reason. The impression this gives students is that atoms are able to judge for themselves what type of bond to make, hence their use of 'want to'. This has become a curriculum model with considerable currency, as students are provided with a convincing reason why chemical reactions occur. At A level, though, this acts as a considerable barrier to learning aspects of chemistry (Taber, 1998b). Therefore, this is a pedagogic learning impediment: something, usually a faulty model obstructing learning, that derives primarily from the usual way science is taught. We suggest that during their exploration of the research literature science teachers identify teaching approaches and ideas that channel student thinking away from the scientific principles we are trying to teach.

LESSON ACTIVITIES IN SCIENCE CLASSES

We suggest above that science lessons are about persuasion and staging the mutual reconstruction of (authentic) scientific models through discourse. Given this, how would we expect science lessons to be characterised? The image suggested is of rooms full of talk, perhaps heated (though respectful) arguments, as evidence is considered and weighed, with students and teacher moving gradually towards consensus. If this is true, then science lessons reflect SCIENCE itself.

Are real science classes like this? In our view they sometimes are, but not as often as we would like. Science lessons commonly include three types of activity that characterise science: talking, practical work and writing. However, these activities are not always undertaken in the most productive way.

The nature of classroom talk in science lessons

Language is a tool for communication commonly used for this purpose in science lessons (Sutton, 1974; Lemke, 1990). A large proportion of the talk is from teacher to the students communicating information and instructions. This is not surprising given that the teacher has more information about the topic than the students, and the teacher is responsible for organising and managing the lesson. However, the teacher can only prepare a lesson so far; once in progress, an interactive process is needed if opportunities for learning are to be optimised (Scott, 1998). So the teacher needs to elicit talk from students, if only to ensure that they understand and follow the teaching.

Language is also a tool for thinking (Vygotsky, 1986). We develop our ideas by 'talking them through'. Mature learners may fully internalise this process, but even the most able and accomplished sometimes benefit from discussing and clarifying their ideas with others. We would therefore expect that lessons that help students learn would include opportunities to use talk to explore and apply the ideas they are being asked to engage with. This could be on a whole class level, but is often better organised around pairs or small group work.

The nature of writing in science lessons

Language is also used in writing. Again, in SCIENCE, writing is both a means of communicating with others and exploring one's own ideas. We hope to see writing activities in lessons as another way of letting students engage with, explore and develop their thinking about scientific ideas.

There is a great deal of writing in many of the science lessons we see. Some of this is designed to provide the teacher with essential opportunities to assess student learning (see Chapter 10). However, much writing we see seems to be primarily concerned with transferring information from the teacher's notes to students' exercise books. Sometimes this is undertaken with some skill, so that the teacher works with the class to build up an explanation of a topic and at key points students make brief notes to record what has been established. This may be an effective way of making notes – brief writing interludes to reinforce the central points deriving from classroom discourse.

Yet, often, the writing takes up significant portions of lesson time, and basically involves copying from a board or screen, or from a book. Given their short concentration spans, and the way some students find careful copying quite demanding, it is likely that much of this copying involves minimum thinking about the material. This reminds us of the definition of a lecture as a means of transferring information from the lecturer's notes to the students' notes without passing through the minds of either (see Table 6.2, p. 129).

Science teachers collectively seem especially concerned that their students should have a full and accurate set of notes, for revision purposes, by the end of examination courses. While we find the teachers' motives admirable, we are less convinced about the need for students to effectively write out a full set of notes. Students should keep notebooks in science lessons as records of their work and to record their ideas, but this should not need to grow into a revision aid. Commercial revision books are likely to be more complete and accurate than the notebooks of most students, and can be purchased at relatively low cost. Student writing activities should engage them in developing their thinking about science. This is more likely to result in recall of the ideas than copying notes. Time is precious, and should be used to encourage thinking and develop understanding rather than to copy.

If students find open-ended writing activities too difficult, their writing can be scaffolded using writing frames (see Chapter 9, and in particular Box 9.1, p. 186) offering varying degrees of structure and support. Teachers can also devise 'DARTs' to help ensure control over the science content is retained while at least requiring some level of information processing by the students (Taber, 2002b).

Types of DART suitable for learning science

DARTs are especially useful in teaching science because of the nature of science 'texts'. In SCIENCE, information is sometimes presented in blocks of prose, but seldom *exclusively* in prose. Scientific texts include a range of other ways of presenting information. These include various types of charts and graphs; schematics; various diagrammatic formalisms such as circuit diagrams, chemical glassware set-ups; chemical formulae; mathematical equations; measurements with numbers, units and implied or explicit precision; photographs and computer-generated images based on a range of inputs.

Many of these types of 'text' are suitable for wide use within science teaching. Tasks asking students to describe the meaning of a trend on a graph, use a passage of text to label a diagram, or to classify data (photographs, drawings, actual samples of chemical, or artefacts) using keys, involve interpreting and translating skills that are central to doing and communicating science. Asking students to undertake these types of activities has a number of advantages over copying. A DART type activity:

- is likely to be more interesting for students and teacher;
- can more readily be differentiated for different students;
- is more authentic as a process reflecting SCIENCE;
- provides a genuine task for students to discuss;
- gives an opportunity to appeal to and develop a range of learning styles;
- is more likely to lead to some deep thinking and so some meaningful learning.

The nature of practical work in science lessons

Practical work is another category of activity common in science lessons. Schools are equipped with teaching laboratories, purchase equipment and materials and employ technicians to facilitate practical work. There is a strong tradition of practical work in UK school science. Most science teachers suggest that science cannot be taught 'properly' without practical work. Many (though not all) students regard practical work as one of the main attractions of science lessons.

If school science is to reflect SCIENCE itself, then we can see a strong argument for practical work remaining a central feature. The sciences do not all collect the same type of data and (as we discussed in Chapter 2) do not all rely on the stereotype of the science *experiment*. All sciences do develop theory in response to empirical evidence and to direct further empirical studies (Lakatos, 1978). Students should, therefore, have the opportunity to experience empirical investigations in the form of 'experiments', as well as field work that gives the flavour of non-experimental (that is, laboratory-based) sciences. We suggested in Chapter 2 that the nature and extent of practical work in many schools has become too closely tied to Sc1 coursework. We believe there is much to be learnt by allowing students to undertake a range of genuine scientific investigations. By 'genuine' we mean that these should not be 'investigations' where the 'answer' could be found by reading the textbook.

However, there is also room for other types of practical work that we would characterise as *demonstrations* rather than experiments. For example, Galileo undertook careful experiments to investigate the rate at which objects fell under gravity. These were genuine experiments, as Galileo did not have access to conclusions in the library; as far as we know, he was the first person to investigate this area carefully. The story of Galileo dropping a canon ball and a musket ball from the tower at Pisa is probably true, but this was not really an experiment to find out what would happen. Rather, Galileo used this as a dramatic demonstration of his ideas, ideas that had already been formed through the earlier research.

Demonstrations illustrate principles we already know. Teachers may use demonstrations to illustrate a scientific principle to students, or may get the student to carry out the practical. We do not think there is anything wrong with students undertaking practical work with known outcomes, as this may help them understand and remember the principles, as well as giving experience of practical techniques. Indeed, there is an argument that when students do not know what the outcome of a practical is meant to be, they commonly misinterpret what does happen (Driver, 1983). With all the effort used in collecting apparatus, setting it up, understanding what to do and doing it while trying to record results there may be little capacity left to think about what happens and what the consequences may be.

We find this argument very convincing. Even knowing that a ball thrown in the air follows a trajectory with constant acceleration does not stop us wanting to imagine that there is a short moment when the ball is no longer ascending, but has not yet started descending. Similarly, no amount of understanding of the mathematical description of simple harmonic motion seems to help some of us 'see' a pendulum bob do anything other than seem to swing at constant speed for most of its cycle.

Asking students to rediscover major scientific principles in school laboratory conditions, with all that entails, would seem to be overly optimistic. We suggest experiments should be genuine and not look to rediscover known information. Asking whether cotton wool or a layer of foam is a more effective insulator for a hot drink provides a real problem (and a familiar context) that makes more sense than trying to rediscover the principle of conservation of momentum. Practical work aimed at reinforcing accepted and known scientific principles is probably better used by first teaching the theory, then getting the students to see how well they can demonstrate the effect.

As SCIENCE is partly a practical activity, we think that students should usually undertake the practical work, the appropriate guidelines having been provided. However, there are some demonstrations better undertaken by the teacher. These can be quite difficult, as they require manipulating apparatus and teaching simultaneously. There are two types of practical work better undertaken by the teacher: where teacher demonstration is indicated for practical reasons and where this is pedagogically more appropriate. The practical reasons that might indicate teacher demonstration are:

- safety – where there are particular hazards;
- difficulty – where particular skills are needed;
- cost – where equipment is expensive and liable to be damaged;
- resources – where there is not enough equipment for the students to undertake the work in sensibly sized groups.

In the last case, incorporating the practical in a 'circus' of activities or dividing the class to undertake two or more activities in rotation would be an alternative.

The second case where we suggest teacher demonstration (at least initially) occurs, is when the scientific principle demonstrated contradicts the students' ideas. The teacher can use the practical activity as a 'critical event', creating an anomaly in students' thinking and, hopefully, inducing 'cognitive dissonance'. For this to be successful the process has to be carefully staged if the students are to attend to, and appreciate the significance of, the outcomes. Sometimes these experiments can be incorporated into a sequence called 'POE' – predict, observe, explain (e.g. White and Gunstone, 1992). The prediction makes students think about how their ideas relate to the scenario, providing a sense of ownership and commitment to the expected outcomes. The observations should trigger the dissonance, providing the impetus for finding an explanation. The scientific model can usually be achieved with the teacher's help.

Before moving on from this topic we advise inductees that a good rationale for using practical work in science lessons is needed. Practical work is included in a lesson for a purpose. Motivating students and giving them enjoyable science lessons is a fine purpose, but as most students seem to enjoy most practical work, this is not sufficient reason for including any particular scientific activity. In Chapter 12, we review research relating to practical work. This indicates that the learning potential arising from these activities is much less significant than many teachers think. So, although practical work can allow practise of technical and investigative skills, and illustrate scientific principles, it is necessary to select the most appropriate practical work to include in a sequence of lessons. This will depend upon the specific learning objectives we want the practical work to help us achieve.

USING INFORMATION AND COMMUNICATIONS TECHNOLOGY (ICT) IN SCIENCE LESSONS

If practical work is a feature of scientific activity, then so is the use of appropriate technology. The progress of science has depended on the development of equipment, for example, the glassware we now take for granted, microscopes and oscilloscopes. In the twenty-first century information and communications technology (ICT) is of particular importance in scientific work. ICT is also regarded as very important in education, with liberal references to it in the NC and in ITT documentation. Those awarded QTS must, for example, demonstrate that they use ICT effectively in their teaching (TTA, 2003a). ICT is sometimes seen to be synonymous with computers, but that is too limiting, as some of our examples below will show. We agree that teachers today should make use of ICT both to teach their subject and to support their wider professional role.

However, as with all areas of professional work, teachers should approach ICT with a critical attitude: make good use of ICT, but there is no need to use ICT for its own sake. We suggest ICT has the potential to help in three different areas we label as administration, educational technology (ET) and authenticity.

Using ICT as an administrative tool

ICT can provide considerable support for the administrative aspects of teachers' work. Some teachers use an electronic mark-book, while others have a lap-top with mark-book, resources and lesson plans linked so that they can move between related documents at the click of a mouse button. There are obvious advantages to using the computer for any documents that need to be communicated, modified over time, or used as templates for related documents in the future. We assume that readers are likely to be 'computer-literate', having used word-processing and possibly spreadsheets, databases and so on. The internet and email have created a revolution in communication, although mainly in ease, access and speed – not necessarily in quality.

One particular feature of the internet worth mentioning is that of electronic discussion lists. Anyone can readily set these up, free, on advertising-supported services and some universities and professional bodies provide lists for the academic community. There are lists for science teachers (see Box 7.2), although some of these are 'high volume' with lots of messages and so once on the list a user must scan the incoming mailings to select those messages that might be of interest and delete the rest. These lists include postings on such matters as useful resources, how to get difficult experiments to work, examination boards and papers, as well as discussing policy issues, subject matter and how to teach it.

Educational technology: using ICT as a teaching tool

The term 'educational technology' was widely used for forms of technology that could support teaching before computers became commonplace in schools and colleges and the terms 'IT' and 'ICT' entered the language. Educational technology (and for that matter ICT) technically included such mundane technologies as chalkboards – or even slates. Educational technologies should be judged in terms of their fitness to help achieve educational aims. One of the current modern technologies is the interactive whiteboard (IWB). This technology has considerable potential as an effective teaching tool, but can also be used as little more than a more expensive chalkboard.

Educational technologies include cassette tapes (for example, of radio programmes), videotapes, DVDs, 35 mm slide projectors, overhead

BOX 7.2 ELECTRONIC DISCUSSION LISTS FOR SCIENCE TEACHERS

Biotutor

> The Biotutor Discussion List is an email Discussion List for biology teachers in schools and colleges, with students up to the age of 18 (in the UK this would include GCSE and A-level), and for those with similar professional interests. The List is intended to be used for the dissemination of good practice, and to assist subscribers in resolving problems related to all aspects of the teaching of the biology in schools and colleges.

> (www.biology4all.com/biotutor.asp,
> accessed 10 April 2004)

Chemistry-teachers

> This list is for A-level and IB Chemistry Teachers.
> (www.uk.groups.yahoo.com/group/Chemistry-Teachers/,
> accessed 10 April 2004)

Learning-science-concepts

> This list is intended as a forum for discussion for teachers, researchers and others who are interested in aspects of learning in science. Particular themes might be: alternative conceptions in science; modelling the learning process; scaffolding learning; developing and critiquing teaching analogies and models.

> (www.uk.groups.yahoo.com/group/learning-science-concepts/,
> accessed 10 April 2004)

Physics teaching notes and comments

> A discussion list open to all with a professional interest in the teaching and learning of physics. The list currently has over 450 subscribers, leading to a reasonable pool of wisdom for tapping into, a sensible receptacle for your tips and tricks, a source of data collection when faced with benchmarking and a useful virtual soapbox for your opinions.

> (www.education.iop.org/Schools/networks/,
> accessed 10 April 2004)

projectors (OHPs) and video cameras. These devices can potentially help the teacher achieve his/her classroom objectives. Introduction of OHPs provided an alternative to using chalkboards, with the considerable advantage that it was possible to write on the OHP film while still facing the class. However, the main advantage of the OHP over the chalkboard was the ability to prepare materials in advance and to project images reproduced from books, journals and magazines. The OHP also permits greater flexibility in use. For example, a clever trick is to use a piece of paper to cover the overhead transparency (OHT) so that the teacher can see the whole (as the paper is translucent under bright light) but the class can only see what the teacher has chosen to reveal. Another advantage is the ability to use overlays – several OHTs that build up a presentation, by filling in answers, or developing a diagram in stages.

Similarly, computers and related technology open up new possibilities. In principle, students can access through the internet an almost unlimited library of material without leaving the room. Also, a great deal of software is available to support science learning, although some is little more than electronic versions of book form materials. Computers allow the possibility to use simulations that were not previously possible and to synthesise presentations including useful visual material. Some processes in SCIENCE are too small, too large, too fast or too dangerous to actually show directly in the classroom, but can be shown easily using computer simulations. The huge processing power of computers makes simulations possible that would be almost impossible in any other way. The advantage over a film animation is that the teacher or student can often change the value of parameters, almost instantly seeing how that process is influenced.

Well-designed simulations can be excellent teaching aids. However, as always, teachers should remember that simulations are computer models of the curriculum models of the SCIENCE, so there is scope for something that looks very impressive, but which is dubious in either its science or pedagogy. Simulations can model SCIENCE poorly, or can simplify the wrong aspects so the presentation is trivial or overcomplicated for students. However, there is educational material being produced where care is being taken over both the science and the educational principles.

Pedagogically informed computer-based learning (CBL): the EPIC model

Computer-based learning (CBL) materials have been developed in the UK for dissemination to the post-16 sector through the National Learning Network (NLN). A company called EPIC designed the physics materials and included the use of animations and interactive screens that engage and interest the user: one of the strengths of the medium. More significantly, EPIC started the design of the materials by considering how they would best help students learn. The physics materials were broken down into units called Learning Objects (LO) each covering a few key learning objectives in the topic. Each LO was designed with the same basic structure (see Table 7.1), planned around a pedagogic model of how the materials facilitate learning. Students can choose to navigate around an LO at will using a menu, allowing for different needs and learning styles, but the recommended sequence of screens follows the pedagogic model shown.

Computer activities need to be chosen carefully with particular goals in mind. Inductees should also consider that although computers are motivational for some students, others find them very boring. Educational

Table 7.1 Structure of an EPIC-designed CBL unit

Sections of unit	Purpose within unit
Introduction	To get the students' attention, and engage them in the unit
Aims	Overview – to show students which physics they will be learning
First thoughts	To activate and start organising the students' prior learning
Presentation	To introduce the key new information to be learnt
Investigation	To take students deeper and further explore the new ideas
Application of knowledge	To encourages students to use new and prior learning
Summary	Summarises the new knowledge
Checking understanding	Allows students to test their new knowledge

software also seldom matches the standards possible in commercially available 'games' where the investment is more likely to make profit for the company. The same reservations can be made about visual material teachers can include in presentations, so it is important to select examples carefully. An enormous number of animations, photographs and video-clips can be downloaded from the internet to incorporate into electronic presentations. Take care to download and use material with copyright waived, or which you have obtained permission to use. Much internet material is offered in this spirit, so will be made available freely for non-commercial teaching purposes.

ICT used for authenticity in science teaching

Science teaching can benefit from considering how computers are used in SCIENCE. If science education is to provide any kind of authentic experience of what SCIENCE is like, then it needs to reflect the way that computers are employed in many science disciplines. Computers are used for data storage and communication, data capture, data interpretation and modelling phenomena. The complexity of some processes makes some of these operations unsuitable for use in schools, but there are certainly ways to use computers to give lessons a more authentically scientific feel. Students, like scientists, can use computers to draft and prepare presentations – for example, undertaking literature searches and reporting the results. Care is needed here that real research skills are taught so that the outcomes are not simply compilations of information 'cut and pasted' from the first few web-pages offered by a search engine (see the comments about the pilot for Twenty First Century Science, p. 57). Similarly, although students often like preparing electronic presentations for their peers, they need to be guided to focus on content at least as much as presentation, otherwise the results may be very colourful and well-animated slides with minimal information content.

Data logging is a major technique for use in modern science lessons. Relatively inexpensive hardware and easy-to-use software allow students to capture data from a range of probes. This technique should be introduced in a simple practical context such as a cooling curve for hot water. The rationale for the approach should be considered carefully. Data logging can give graphing in real time, with the option of easily changing scaling to see the effect on the graph. This can be much more effective than manual graph-plotting (which also has its value and place in science teaching). The power of data logging is the ability to collect data in ways that are either impossible or extremely difficult manually, such as from several channels instantaneously; at levels of accuracy not possible with taking readings by sight; or using data-capture rates of many readings per second, or over periods of many hours.

Computer modelling in science can require very advanced skills, but there are now applications available that allow students to use computer-modelling techniques. This is different from computer simulations discussed above, when the student can only change the value of parameters that are already part of a given model. The IoP-sponsored *Advancing Physics* course (see pp. 84–6) includes modelling software that students are invited to use to undertake their own modelling of phenomena.

FROM AIMS AND OBJECTIVES TO DECISION MAKING

In this chapter we have looked at some aspects of planning to teach the science in the curriculum. In particular, we have explored how the nature of SCIENCE invites us to select lesson activities. We argue that in our selection from these various activities – be they talking or writing activities, practical work or computer-based – teachers should think about how the activities relate to what is fundamental and characteristic about science as a school subject. We have also made the point that activities are included in lessons to meet specific needs. There should always be a rationale for the inclusion of a particular activity, in terms of meeting the lesson objectives.

THINKING ABOUT PRACTICE

- Use the NC outline (see Table 2.1, pp. 22–3) to find one topic in which you have strong scientific knowledge. List the key scientific ideas you would wish to communicate to students about this topic. Find out which ideas students find difficult and list any alternative conceptions. Discuss with colleagues how to address these so that the scientific ideas are made clear to students. Think about activities you may do and how you would explain the ideas.
- Now use the NC outline to find one topic in which you do not have strong scientific knowledge. Find out the key scientific ideas you may

need to teach and list these. As with the first question, find out which ideas students find difficult, together with any alternative conceptions. Do you hold any of these alternative conceptions? (Be honest!) Work with an 'expert' colleague to sort out these so that you can explain the topic confidently and using accurate science.

■ Which of the key points about learning in Table 6.3 are relevant to the ideas that have been discussed in this chapter?

Chapter 8

Acting to teach science

I love acting. It is so much more real than life.
(Oscar Wilde)

The best actors do not let the wheels show.
(Milan Kundera)

It beats working for a living.
(Anthony Hopkins)

Teaching is a form of acting. To get our meaning, try substituting 'teaching' or 'teachers' in the quotations above. (The last is rather 'tongue in cheek'.) Every day, teachers 'perform' in front of their audiences, with the aim of making or initiating some change/changes in the audiences' thinking about whatever subject is being 'acted out'. Ways in which science teachers do this are the subject of this chapter. Many books combine the words 'teaching' and 'science' in their titles, so in one chapter we cannot pretend to cover large amounts of ground. Questions about this may, therefore, remain unanswered. The views expressed here are very much based on our own experiences and preferences for the act of teaching science.

In the first section, we unpick the processes involved in acting to teach science in the classroom/laboratory. Then we explore more closely the 'pedagogic content knowledge' (PCK), that is, the tools science teachers need to ensure learning takes place.

ACTING TO TEACH SCIENCE

Effective and non-effective 'acts'

Standing up in front of any audience requires some sort of change in behaviour from being a 'normal' person to an 'actor'. Teaching is a job that requires this on a daily basis, with the added demand of not providing 'entertainment', but that the audience learns something from the teacher about science. We illustrate this by providing two accounts of experienced teachers acting to teach science. Read the accounts in Box 8.1 and consider your answers to the questions.

The incidents in Box 8.1 are true stories that happened to one of the authors. They illustrate well the point about acting to teach science. Both of the teachers had plans for these lessons – Mr Y clearly was intending to impart some *formal knowledge* about radioactivity, using dictation as a strategy. Dr A, on the other hand, was doing something as an end of term *entertainment*, which although effective and informative, went awry. Both can be seen as 'actors' – Mr Y speaking in a serious, probably rather pompous way using technical language, adopting very formal postures and habits which leave him standing outside the children's world; Dr A being rather more informal, but characterful and commanding in his manner – the memories 'looking around the room sharply' and 'Here, catch a whiff of this!' indicate a teacher interacting with the class and involving them in events, such that learning felt more of a shared experience even though the children hardly dared to speak to him. The teachers' responses to the incidents also tell us something. In both cases, the class (mainly the same children) ended up laughing, but for quite different reasons – Mr Y's ignorance of the increasing disruption around him and frozen embarrassment meant the children laughed *at* him, while Dr A's cool 'Oh, dear!' with his realisation that the children knew the window was closed *before* he lit the gas mixture meant the laughter became *shared* between the teacher and children. Notice, too, how the memories of the teachers' appearances, dress and style have remained powerful nearly 30 years later and also how critically the children looked at the teachers at the time. Clearly, Dr A's act was far more effective in teaching science than Mr Y's, facts identified very easily by the children in their audiences. The result was that Mr Y was not taken seriously, while Dr A was revered and respected. The difference continued into the examination results too – the same children performed much better in chemistry O level than in physics.

BOX 8.1 TWO TEACHERS ACTING TO TEACH SCIENCE

Scene 1: A comprehensive school in rural England, 1977; a 4th form (year 10) O level physics lesson in the early afternoon on a hot summer's day.

A student in the class remembers:

Our teacher, Mr Y, was walking around the classroom dictating notes to us, I think it was about radioactivity. Physics with Mr Y was always boring. His experiments never worked, but he insisted they worked five minutes before we arrived. He made loads of mistakes doing calculations, which some of us loved to notice and put right. He looked weird with long, blond hair and scruffy beard. In one lesson Mr Y was wandering around talking very seriously into the floor, dictating notes to us about radioactivity or something. After a while someone noticed that the zip of his purple crimplene trousers was wide open and we could see his pants underneath. This went round the class like a huge whisper so soon all anyone could think about was Mr Y's purple crimplene trousers and his zip being undone. He took no notice at all and carried on going round and round talking until we could hardly keep ourselves from laughing. After a few more minutes, he went to the front bench. He put his hands a shoulder width apart on the bench and leaned forwards as he was talking. He did this a few times until one of the boys slid a note under his nose as he leaned forward. The note said, 'Sir, your flys are undone.' Mr Y saw the note, went bright red and dashed out of the room, and we all burst out laughing like a balloon had popped.

Scene 2: The same rural, English comprehensive school, December 1977, last 5th form (year 11) chemistry lesson before Christmas.

The same student remembers:

Our chemistry teacher, Dr A, looked like a pixie. He had startling blue eyes that looked at you in a piercing way over black-rimmed spectacles. He had bushy white hair standing up from his head and a white beard. He was quite short. He had some funny habits, like sitting on the bench with his legs crossed looking around the room sharply, trying to catch us out doing something we shouldn't. This made him seem even more like a pixie. He also liked doing experiments on the front bench making smelly gases. When he made a new gas, he would shove the test tube under the nose of the nearest child, saying, 'Here, catch a whiff of this!' The child would cough and splutter. One of the best things about Dr A was his end of term experiments. He always ended the term with a demonstration. On this day, he was firing a hydrogen-oxygen rocket. He set up a plastic bottle at an angle in a clamp stand and filled it with hydrogen and oxygen from two cylinders then plugged the top with a cork. He talked all the time about what he was doing. We sat on the benches around watching. We did not speak. He got the bottle ready and was preparing it really carefully. 'We're going to fire it out of the window to hit the music block next door!' he said. At the very last minute when Dr A was about to take away the cork and set off the rocket we all looked at the window – it was firmly closed. Somehow none of us dared to tell him, or perhaps we just wanted to see what would happen. One second later there was a huge explosion of the rocket going off and the window shattering on to the ground outside. 'Oh, dear!' he said. Everyone laughed so hard and then he realised that we knew the window was still closed. 'You mean lot', he said, with a smile.

For some inductees learning the act of teaching is often the most challenging aspect of taking on the new professional identity. Some new entrants have had some prior experience of 'acting', perhaps as 'explainers' in a museum, as a medical sales representative or as a youth leader in church and secular settings – or sometimes even an unqualified teacher working abroad. For most inductees the notion of appearing in front of a critical teenage audience, having to control their behaviour and explain science to them is new and, understandably, rather intimidating. Nerves are very common in the early stages, but most inductees have the inner confidence, determination and skills to work through these and ensure that they make a positive start. To help, we begin by looking at the characteristics of an effective 'act' using the two episodes in Box 8.1 as a guide.

THE CHARACTERISTICS OF A GOOD 'ACT'

We can use the two scenarios in Box 8.1 to unpick some characteristics that make a good 'act'. You will undoubtedly be able to add some of your own, or identify with these based on your own experiences of school and/or 'acting' in a specific role. A useful activity may be to recall similar episodes from your own education and reflect on these in the same way. We give our characteristics as a list, discussing each in turn briefly.

Good subject knowledge The accounts note that Mr Y's experiments had a reputation for failing. The author recalls many lessons where students were expected to sit still and watch, only for nothing to happen. She also recalls trying to learn Newton's laws of motion from this teacher while he wrote and re-wrote calculations on the board due to a large number of errors. This resulted in a poor level of student confidence in the teacher. In contrast, Dr A always appeared to be in command of his subject knowledge, seemingly always confident and accurate (the 'appeared to be' and 'seemingly' are important). His experiments always produced interesting (if dramatic) results and they were executed in an assured, confident manner that held attention. These two teachers illustrate almost perfectly the old aphorism: 'If it moves, it's biology, if it stinks, it's chemistry and if it doesn't work it's physics.'

Interacting with the students Mr Y appeared aloof, pompous and lacking in humour. His actions did little to stimulate the students to empathise with him in any sense. His stance was to remain distant, even when the class was obviously disruptive, as if this was how he perceived the 'act' of teaching should be. Somehow, the students respected Dr A far more. He was able to be distant and formal at the same time as being observant and engaging. He captured students' attention, but also made them think he could see everything they did.

Making learning happen 'Learning' in the 1970s in the author's school involved quite a lot of rather boring moments, so poor Mr Y's strategies were probably not unusual. However, there was a difference between these two teachers, in that Dr A had the knack of entertaining when it mattered. At a basic level too, Mr Y probably saw teaching as imparting knowledge into blank slates (cf. Table 6.2). Dr A made it more of a shared experience by using 'We' rather than 'I', being aware of students who did not understand first time, and maintaining involvement by always talking through what he was doing, even if it was only him doing it. Mr Y's physics passed before the students' eyes as a mysterious charade. Dr A forced students' participation, even if his strategies in reality were probably little different.

Appropriate dress and personal style Mr Y is described as looking 'weird' and his taste in clothes is noted; Dr A had a characterful appearance that he carried off well and his clothes must have been neutral as these made no impression. Some students are great observers of teacher fashions, hair, earrings, hand movements, other mannerisms and habits.

Position Although Mr Y walked around the lab, he looked at the floor, not at the students. Dr A looked at the students 'in a piercing way', making them *feel* everything they did was noticed, even though he often sat still.

These two teachers were very much products of the 1970s era. To be fair to them both, and reflecting back now as a professional teacher, it is easy now to see that Dr A was a former grammar school teacher (see Chapter 3) with years of experience, while Mr Y was a much younger, trendier product of the 1970s (for example, wearing crimplene, a fashionable material at the time), possibly in his first post. Dr A's whole act was inevitably more polished than that of his colleague. The critical point, though, is that the students they both taught were oblivious to any differences of experience – the cruel reality is that criticism is levelled on the same criteria at every teacher who stands in front of them, making no allowances whatsoever. Therefore, for the inductee, the vital question is, how can a person with little or no prior experience turn into an effective 'actor' for science? What can a trainee teacher do to avoid a Mr Y scenario? Another way of thinking about this is to ask 'what qualities must be acquired and are any innate'? Is it true that 'good teachers are born and not made'? Some qualities must be innate, perhaps for example, a profound interest in communicating a subject, a generally outgoing personality and a willingness to work with young people. Yet there is still a lot that can be *learned* about developing the *craft* of teaching – if you consider training to be a teacher rather like going to 'acting school' then you will not be far off the mark. So, let us begin by continuing the analogy to acting; what preparation might a (science) teacher do in order to have successful encounters with a very critical audience?

WHAT COMPRISES ACTING TO TEACH SCIENCE?

Here we give some practical advice based around our own experiences of teaching and helping inductees. We also give you the British Army adage of the 5Ps: 'Poor planning produces poor performance.' Good planning is the essence behind a professional teacher's work and is at the heart of making a difficult job appear easy, giving the students (and teacher) confidence and making best use of the available resources. Here, we offer practical advice about how to plan your teaching 'act'.

Learning the script: learning and preparing your material

At a basic level, a teacher must know the material s/he is to teach, or at least, be able to appear confident enough to hold the attention and answer the questions of a potentially critical audience. Beware of 'appearing to be confident' – students have an uncanny ability to tell when a teacher is bluffing and when s/he really 'knows their stuff'. Those who are bluffing, and are found out, can get very rough treatment.

In the early stages of preparing lesson material, our advice is to try to teach science that you know well. Do not commit yourself to explaining anything in detail that you are too uncertain about (although this may be outside your control). To test for level of uncertainty, try explaining what you are about to teach to a colleague, partner or critical friend. Ask for feedback on your explanation and get them to ask you questions once you have finished, perhaps as if they were the students. If your explanation is unclear or too complicated and/or you feel extremely uncomfortable with the questions and cannot answer them, this is a clear indication that you are not really that certain of the material you are planning to teach. If the topic is outside your area of expertise, watch out for alternative conceptions you may have (see Chapter 5), sort these out by discussing your thinking with an expert and/or read through textbooks and research literature (see Thinking about practice, pp. 156–7). Honesty with yourself is definitely the best route, because if you are 'found out' to be bluffing by a class they can make your life more difficult for you than a caring friend will. If you find weaknesses, you must go back to the material and work on it some more before explaining it to a group of hormonally challenged teenagers.

The practical strategy of using cue cards can help maintain the flow of a lesson, particularly if the topic is unfamiliar, the class especially challenging, or if you are prone to get very nervous. Preparing these takes time, so it is probably not a realistic strategy for every lesson, but the technique is useful. When one of the authors gives demonstration lectures to large audiences, she prepares cue cards to help maintain the lecture should something go awry. Information is written in short sentences, in large, clear letters on A5-sized lined cards, on one side only. Each card is numbered. Even instructions for the experiments are written in this way and included in the sequence. Notes are included indicating which overhead transparency to use and/or item to show at any point. Holes are punched in a top corner and a tag placed through them all so that the cards can be read in sequence and turned easily to maintain a neat pile. Even if the cards end up not being used, it gives a feeling of security that if something unexpected happens, the thread of the lecture can be picked up without difficulty, or, if nerves strike, then the story to be told can still be followed. Cards look neater than a sheaf of closely written A4 paper, which tends to flop and quickly become scruffy. The author prepares the cards, at the same time memorises the lecture, then sits and reads through them for a few minutes about an hour before going 'on stage'. The cards are there either to be used, or to act as a prompt in case she gets stuck. To help with demonstration experiments, individual instruction cards are left in trays with the chemicals to ensure that the demonstrations, at least, can be done correctly.

Take advantage of any ICT available, such as IWBs (see p. 151). These can shorten preparation time and allow the creation of resource material that can help you bring a topic alive. We discuss what sort of activities to prepare below.

You can see from this that not only does an inductee need to know and understand the science, but the material has to be prepared in such a way that the lesson can run smoothly. We are not suggesting that each lesson is a lecture, but that whatever is done in front of the audience appears to be done without letting 'the wheels show', as one of the opening quotes states. Knowing that materials are well prepared and anything you are planning to explain is known to you in a way that you can stick to when under pressure is one aspect of giving yourself confidence. With time and experience, preparation of subject matter becomes easier and more routine, so cards or other forms of notes become less necessary. However, it is vital to prepare thoroughly when teaching a topic for the first time, however experienced you become.

Another aspect of learning the 'script' is to *pay attention to detail and not to leave anything to chance*. These factors must always be attended to and can never be substituted for by 'experience'. Part of being a professional teacher is to ensure that sloppy habits do not creep into daily practice. Classic examples include: not testing an experiment before either a class of students does it, or before you have to demonstrate it your-self; finding yourself stranded in a distant room or laboratory trying to do an experiment, only to discover that a key piece of equipment is missing because you did not check it was there beforehand; remembering that the 'last time I did this' the matches were in a certain place, only to discover this time they are not; and doing a complex calculation on the board without knowing the answer. All these situations are poten-tially extremely embarrassing, resulting in considerable 'loss of face'. For an inductee, building 'credit' with students helps develop a secure feeling between you and them, so that they can rely on your knowledge and expertise to help them learn. All these scenarios have the potential to destroy something you are trying so carefully to build.

Experienced science teachers with substantial 'credit' already in place find they can live through such occasional situations, because the students know these are not usual. If they are usual, students respond as we did to Mr Y. Students can also, with an experienced practitioner, work with him/her to find solutions to problems or solve practical difficulties. Inductees do not have this luxury, so our advice is not to take the chance. Finally, try to keep your own personal working area in a laboratory clean and tidy, as this also creates an organised and professional appear-ance. If you are habitually untidy, try to keep this 'behind the scenes'. Another aspect of this self-organisation is to take great care of any work that students hand to you – they become very frustrated if this is lost in a pile of paper, chewed by your own dog (rather than theirs) or used to soak up spilled coffee and then returned after several weeks in bad condition.

Knowing your audience

The next vital point is to know your audience. Of course, the first time an inductee takes a class this is impossible, but try to be aware at least, of what you are facing. The first few lessons give opportunities to get to know your students by name, which is vital. Learning names can be challenging for some inductees. A variety of strategies can be used for this, including:

- Making a seating plan of each class showing where students sit – this can work, provided the students don't move before you have got to know them.

- Using school photos to identify students – have these in front of you when teaching.

- Using a note system – for example, making short, identifiable descriptions of students against a list of names, for example, 'tall, long ginger hair, tied back, glasses'. This is all right provided the descriptions are not rude and will not be seen by the students.

- Having the strategy of always asking students to give their names when talking to you for the first few lessons and working hard to use these in responding. A technique used by politicians is to look hard in the person's eyes as they say their name and then to use the name in the first sentence you say back to them. This is supposed to help the mind attach the face and name.

Bear in mind that even as an inductee you may have several hundred names to learn. If this kind of memorising is a challenge for you, then you must work at it. One of the surest ways of gaining the attention of a student is to use their name – some students will be pleased to avoid your knowing their names as long as possible, for their own convenience.

There is, of course, another sense in which as a professional science teacher you need to know your audience. This is as a group of learners. Every student will have his/her set of preferred learning styles and habits. Here, we offer two ways of looking at learning styles and habits. First, the educator Guy Claxton (1990) describes possible student 'stances', by which he means the multifaceted package of strategies students have for dealing with lessons. (Note that students are also 'actors' on the stage of the classroom or teaching laboratory.) These can be summarised as follows:

The Swot: a compliant student who generally stays out of trouble and prefers to be doing 'proper learning'. This type prefers copying tasks to being made to think.

The Boffin: a student who absorbs intellectual knowledge of interest to him/herself. This type is often very selective about what interests him/her to the exclusion of most other topics or subjects.

The Socialite: a student who uses the same kinds of strategies to pick up information in lessons as when talking to friends, watching TV or reading magazines. This type may be chatty and easily distracted, interested in personalities just as much as ideas.

The Dreamer: a student who is often absorbed in an inner, fantasy world and seeks not to be noticed by sitting in the spot in the classroom where s/he is least likely to attract attention. Hence the term 'radiator kid', as sitting next to the radiator is a quiet place.

The Rebel: a student who seeks approval from other students by being a joker or more of a 'hardnut'. 'Very able' students who adopt this stance can be the most challenging to deal with (Claxton, 1990: 123–9).

A second way of looking at students is provided by Howard Gardner, whose theory of multiple intelligences (Gardner, 1993) has gained popularity in recent years. Gardner proposes that there are at least seven intelligences possessed by most people, with individuals differing in the degree of skill and nature of their own particular combination. He defines an 'intelligence' as a set of mental skills which enable a person to solve a problem or make a product that has meaning in a specific cultural setting. The intelligence must also be codeable, that is, it can be represented in a symbol system, such as musical notes, formulae, language or pictures. Gardner's original seven intelligences are:

- musical, that is, having musical ability;
- bodily-kinaesthetic, meaning awareness of one's own body movement;
- logical-mathematical, which involves solving 'scientific thinking'-type problems;
- linguistic, showing strong awareness of grammatical structures, use of words and language;
- spatial, used for navigation and reading maps;
- interpersonal, which means having awareness of other people's emotions, moods and intentions; and
- intrapersonal, that specifies internal awareness of one's own feelings and moods to guide behaviour.

Gardner (1999) has since considered that there may be other category intelligences, including – of interest to science teachers – that of naturalist. However, the list of seven intelligences given above is the most commonly discussed.

A related notion is that of 'learning styles', which refers to the 'preferred' way in which someone learns. For example, some students seem to learn better by reading, while others may prefer to listen (and may perhaps seem to be staring into distant space as they concentrate on what they hear). For other students, so-called kinaesthetic learners, physical movement may be important – and learning effectively while keeping totally still is actually a challenge. Although there has been some very useful work on learning style (e.g. Riding and Rayner, 1998), it is still an area that is generally not well understood within teaching (LSRC, 2004).

Gardner's views suggest that a class of students will include a variety of types of learner in your science lessons, of whom only a small proportion are likely to have strong logical-mathematical intelligence, the most 'scientific' of the seven. Similarly, students will vary in the ease with which they make sense of information presented in different modalities (i.e. talking, writing, musically, graphically). This presents quite a challenge for the teacher as science lessons should be approachable and accessible to all – this requires having varied activities reflecting students' different strengths and learning preferences (see Table 6.3, p. 131).

The costume: preparing your appearance and dress

Actors need the right costume on stage – these are used to create dramatic effect in whatever way the director chooses. Teachers, too, are 'on stage' – their audiences will also notice their 'costumes'. Teachers, of course, have varying styles of dress and many schools have their own 'dress codes' which we strongly advise all inductees to follow. Whatever you wear, be prepared to be seen. By this we mean that the students will *notice* whatever it is you wear. An English teacher colleague of one of the authors always wore black. She took this as her 'dress code', because, as she saw it, this avoided comments from the students about what she wore. However, the students still commented privately that Ms X 'always wears black' and that is just 'so boring'. Male teachers in many schools are expected to wear shirt and tie – and, of course, the ties and shirts come in for comment from the students. Whatever a teacher wears may be noticed, even down to tiny details like earrings, rings on fingers, shoes, belts, other jewellery, tattoos and hair fastenings. In general, students prefer to see well-dressed, professional-looking teachers to those who appear 'weird', 'strange' or 'scruffy'. Despite their own dress sense, students are surprisingly conservative. However, this is not to say that inductees should shred all traces of individuality and become clones – students also respect teachers being individuals, characterful and interesting. One author used to enjoy wearing bright colours when teaching in a school with a very dull-coloured uniform, simply to enliven the room and create an impact, but changed to wearing dark colours when needing to 'hide' as a classroom observer. The key point is for inductees to follow whatever dress code there is in the school in which they work and to be prepared to 'stand up and be counted' for whatever they do wear in front of their classes.

One issue for science teachers is that of wearing a lab-coat or not. Our view would be that, in general, this is a good item to wear when the lesson is likely to involve making – what could be termed – 'a mess', that is handling chemicals or materials of any kind that could damage your own clothing. If one is worn, this should be done up down the front and not left flapping open, which not only looks sloppy, but is also pointless as an open coat offers no protection (and can even catch on apparatus and more readily get dipped into solutions). Along with this, science teachers must also set good examples in all other aspects of safety, including tying back long hair, making sure ties and scarves do not flop around and always wearing safety glasses when necessary. For science teachers wearing ties, the use of a tie clip or tie pin avoids having to tuck the tie into the shirt. One of the authors uses a tie clip habitually, but usually clips the tie to the shirt using the label at the back of the tie – this does the job well without the clip being visible, which might look too 'showy' when working with classes in school.

The authors have frequently been surprised at how little attention is given by trainees to these points. Students can get used to teachers wearing lab-coats and there is a stereotypical view of a 'scientist' which many hold – that of a lab-coat-wearing individual (usually male) wearing glasses/safety glasses and mixing chemicals. Wearing a lab-coat somehow 'looks scientific' to students, which may be a good thing or not, depending on their view of science. We would advise inductees to wear a lab-coat when necessary, as this is good safety practice.

The 'teacher' voice

As for actors and opera singers, the voice is the most important tool a teacher has. An effective teacher will be able to use his/her voice in a multitude of ways – to explain, condone, cajole, joke or reprimand, as well as modulating the pitch between a whisper and a very occasional shout.

Teachers use their voices continually, so care must be taken to protect this vital asset. A good starting point is to learn to project the voice across a room using the diaphragm, rather than straining the vocal chords. The trick first involves breathing properly, feeling the diaphragm moving and then realising that the voice can be projected without straining the larynx. If necessary, take advice from a drama professional, working with him/her to help you get this right. The practice and any expense are well worth the effort, as your voice will be your main working instrument for your teaching career.

Other issues about speaking well are valuable for the audience. Words need to be spoken clearly, not mumbled or slurred together and should be pronounced correctly. Accents and dialects do not matter – but be prepared if you do have a pronounced accent and work in an area of the UK which has a different one that it will take time for you and the students to get used to each others' ways of talking. One author found that about six weeks were required to adjust to hearing a new accent among the students on moving to a different part of the UK to teach. During this period, she spent the entire time asking people to repeat what they said sometimes several times, while she was regarded as being 'posh' because she quite clearly did not come from that area. Accents and dialects often add character to voices, making them easy and pleasant to listen to, so, provided the speech is itself clear, this makes no difference to the impact on the students.

Try to avoid linguistic habits such as 'um', 'right', 'OK', 'there we are' or whatever. These tend to get a teacher 'known' and any clever mimic will be very willing to imitate your manner of speech when the moment is ripe. Ask your kindly critical friend to listen to you speaking and to listen for any linguistic glitches you may have. Back in the author's 1970s comprehensive school we spent many happy hours counting the number of times one teacher said 'um' in a lesson, reaching a record of over 80 in one hour.

Learn to use the voice effectively – this means also not using it too much. We mentioned before that inductees tend to talk more than is necessary in their early lessons – the voice will certainly be worked very hard if lessons are just lectures. Try to restrict the amount of whole class talk required by ensuring that students have plenty of activities to do. Supporting learners while they are working on tasks requires different voice skills to talking to whole groups.

171

The stage: moving and standing appropriately and using props

As a teacher, moving around the room and using the classroom or laboratory space effectively gives a powerful dimension to your act and can help significantly with maintaining classroom discipline. Making sure that any 'props' you have, such as activities for the students, items to show, or experiments to demonstrate, are organised to be readily available, also helps create a professional impression, reducing the opportunity for a class to become fidgety or disruptive.

We recommend that inductees do not get into the habit of standing in one place and using every lesson to let their own voice be heard for long periods, as if declaiming Shakespeare's monologues. However, all teachers do have to get control of the whole student group at various times and stand in a relaxed way explaining, assisting with a transition from one activity to another, giving instructions or reprimands. Actors are trained to move and stand on stage so that their movements seem natural and unconscious, rather than strained and stiff. Inductees unused to performing in front of an audience may feel awkward, unsure what to do with their hands, aware of feeling 'rooted to the spot'. Try to relieve nervous stiffness in your body resulting from tension by breathing deeply and forcing your shoulders to relax. Putting a piece of chalk or a pen in one hand will ensure this hand has an activity. Try not to become a habitual 'hand-waver', as this can be very distracting. Place the focus of the lesson firmly on what the students do – this is part of organising the 'props' needed for your lessons to run smoothly. Not only is this good pedagogical practice, but attention is then removed from you, as the students have a task on which to focus.

Preparing the activities includes preparing materials for the students to work from. These, as we will see, can take various forms. When the students are engaged on an activity, their focus will be shifted. This should automatically help you relax and you will be freer to move around the room. When moving around, the best piece of advice we can give you is to always position yourself such that you can see as much of the room and therefore as many of the students as possible. This gives the opportunity to develop good observational skills by scanning the room regularly, so any trouble can be stopped at an early stage. Inductees often find this skill very difficult to acquire, as students can be very engaging and tend to 'capture' you into long discussions, leaving the rest of the group unattended. We have seen many inductees with their heads buried in an intense discussion with one or two students while the rest rove unchecked about the room. Try to avoid this by keeping comments to students short; the art of 'moving on' is useful. One way of developing this is to set the target of talking to every student in the lesson. This will make it impossible to spend too long with any one person or group.

If the class is working in silence or very quietly on an activity facing towards the front, then it is often a good strategy to position yourself where they cannot see you, e.g. at the back. The first student who looks around to find out where you are is probably a 'rebel' (see p. 168) looking to create something other than a work-related situation.

The opening night: preparing the critical first meeting

The story is told of a trainee who observed an experienced colleague taking a class he would be teaching for the first time the next day on his teaching practice. The experienced teacher talked for a few minutes at the start of his lesson, explaining what he wished the students to do. He then said something like, 'Right, you know what to do – off you go, then!' with the result that the students went in an orderly fashion to collect equipment, set it up and perform the experiment. The next day the trainee mimicked the act he had seen the day before, explaining a little about what he wanted the students to do, then saying almost exactly the same phrase as his experienced mentor. The result was chaos – screaming students out of control. Why did this happen?

The answer lies in the fact that students perceive each teacher individually, being quite incapable of transferring the world created by one teacher to another. In other words, it is up to each teacher to create and perform his/her act – and they will respond accordingly. In the above account, the experienced teacher had, over time, worked with the class so that when he said the phrase, 'Off you go, then!' this had evolved into the signal for the students to undertake a particular set of actions *agreed with the teacher* in advance. The exact phrase coming from the trainee meant nothing, because he had not made such an agreement.

As this story illustrates, the first meeting with a class is often regarded as a critical event. Experienced teachers and inductees vary in their ways of approaching this (Wragg and Wood, 1994). Two contrasting episodes from the films *Mona Lisa Smile* and *Dead Poets Society* illustrate this point. Both these films are about educational settings, showing how a teacher impacts upon the students s/he comes into contact with. In *Mona Lisa Smile*, set in 1950s' America, the young, inexperienced teacher (played by Julia Roberts) is shown giving her first lecture to a class of clever, articulate female students at the all-women college at which she has dreamt about working. The lecturer has strong ideals and beliefs and thinks she has prepared thoroughly. Her first lecture turns out disastrously, as she has seriously underestimated her students' prior knowledge. They treat her very badly, pouring such scorn on her efforts that she leaves the lecture room in tears. She recovers, of course, by reflecting on where she went wrong, taking the challenge of working with these girls so that in the end she is well respected.

Dead Poets Society is set in a boys' boarding school, also in America. Robin Williams plays an experienced English teacher whose first lesson with his students takes them all by surprise. The film opens by showing the first day of term and tiny clips of the spectrum of teachers working in the school. Most follow conventional patterns of behaviour, reciting tasks and giving dictation. Robin Williams's character starts by entering through an unexpected door, whistling through the room and walking out of the other door, astonishing the students. He beckons them to join him in the hallway where he meets them for the first time in front of school memorabilia. A discussion ensues in which he is not only able to discern something about many of students, but also indicates to them what he is expecting.

If you know the films, you will also know that the outcomes for these two teachers and groups of students are very different. We simply use these as examples of meeting a class for the first time. The young lecturer in *Mona Lisa Smile* suffered a major setback after her first lecture, while the English teacher's first meeting could be said to be a success. What made the difference? On meeting new students, experienced teachers will tell you:

- to appear clearly in charge;
- have an activity ready which you know they can do;
- project a firm image, perhaps of mystery (but don't overdo it);
- establish rules for behaviour which everyone can adhere to;
- keep your distance and do not appear too friendly.

The lecturer in *Mona Lisa Smile* did not really take charge of her class, but allowed them to dominate her. In contrast, the teacher in *Dead Poets Society* had the confidence to act to create surprise, placing the onus on the students to respond to him and deflecting any plans the students may have had to cause disruption. One of the authors heard of a science teacher who met new year 7 students with a special activity to create a surprise. This involved a kind of chemical 'feely box': a black box with various tubes coming out of it into which she poured different liquids to see what would happen. The author does not know the full details, but when a liquid was poured in, another liquid of a different colour came out from another place. The students were challenged to think about this as a scientific problem they had to solve, the activity being used to introduce what, for this teacher, SCIENCE was about.

The basic principles for a first encounter, then, are:

- to have a very well-prepared script;
- to have good activities that capture attention;
- to set expectations for behaviour and to reinforce these;
- to set a professional stance between you and the students, communicating that you are in charge as their teacher, not their friend;
- to do your best to appear relaxed and in control, however nervous you may feel.

Usually, first lessons go much better than inductees expect. Students are normally very wary of new teachers, so tend to give them the benefit of the doubt until they know otherwise. This is the beginning of what is often called 'the honeymoon period', which can be long or short, depending on a range of factors. Therefore, in a way, it is not really the first lesson that counts for the students – they don't know you, or what your expectations are. The point is to set these out in the first lesson and then to reinforce them in subsequent lessons, establishing a working relationship that creates a solid, peaceful and pleasant environment in which learning can occur.

If and when things do go 'wrong', then you have to work your way back. In this situation, you will need support, encouragement and advice from colleagues. Here is an account of one of the authors experience of meeting a year 9 science class for the first time on her teaching practice, which took place in a comprehensive school in north-west London in the 1980s:

This class had a notorious reputation in the school so perhaps unsurprisingly I felt apprehensive of taking this group for the first time.

My first lesson with them was the last lesson on a Thursday in my second week in the school. Quite good experiences earlier in the day teaching for the first time had boosted my confidence. However, the lesson with this class was simply a disaster. The students ended up running around the room making huge amounts of noise and not doing at all what I had planned. I felt completely powerless to stop them and neither could I see what I had done wrong. The Head of Science, their usual teacher, watched the debacle from a discreet position at the back, where he was marking books. He commented afterwards, 'I hope that is the worst lesson I will ever see you teach.' Towards the end of the lesson, another teacher of the class came into the room. This completed my humiliation, as the moment they saw her the students became completely silent and stood absolutely still. She spoke only a few words and they ended up sat in silence. The next day I talked this over with my tutor. I asked him how she did it. His reply was, 'One day, you will know.'

The story does have a happy ending. After taking advice, preparing more thoroughly and working to get to know the students, the situation changed over a period of time, such that after about six weeks the students were working quite well and by the end of the practice about three months later good control had been achieved, so the students seemed very pleasant and hard-working. Most trainees or inductees will at some time have a similar experience to that described above. To resolve this kind of situation there is no substitute for hard work. The secret is to be determined, to work with advice and above all to make science interesting and lively for the students.

'IT BEATS WORKING FOR A LIVING': TEACHING AS A VOCATION, NOT JUST A JOB

Although the last quote with which we began the chapter was included in a 'tongue in cheek' sense, that is, not entirely serious, there is some truth there, as for many teachers teaching is a vocation, not just a job. Working with young people can be extremely challenging, but is often as equally rewarding, even if not in financial terms. In the final part of this chapter, we reflect on other aspects of the act of teaching that can make all the difference between the hard effort and the sense of reward.

Perhaps one of the first aspects is that inductees make a shift in perspective. Initially, the focus for many trainees on teaching practice is their own 'survival' – that is, 'What can I do to get through this experience?' This is often the case for the first few lessons – establishing the feeling of working with young people in teacher 'mode', preparing material to teach, and getting the sense of classroom management, are all necessary. However, this can and should only last for a limited period, as the most important part of any lesson is *what the students do*. Taking this as the focus means the perspective shifts from 'survival' to 'learning'. This is, to borrow a phrase from government, a 'key stage' in becoming a professional science teacher.

As trainees gradually become more accustomed to being in the teacher role, other aspects of the job become more apparent. Experienced teachers often talk about having a variety of roles with their students, including counsellor, guide, diplomat, social worker and general handyman. The sense of involvement with young people can become very intense, especially if a teacher takes on a pastoral role within the school as well as being a subject specialist. Having time for oneself, therefore, becomes important and necessary, otherwise a trainee can easily become subsumed by learning the teaching role, completing coursework and doing preparation, but also sense a loss of 'self'. Suddenly, you become 'a teacher', not 'a person'. A former colleague of one of the authors told how a girl (aged about 11) he taught was out shopping with her mother when she saw him in the same shop. 'Look, mummy,' she said, 'there's Mr X! I didn't know he was a *real person*!' It is surprising how strongly teachers are identified *as teachers*, losing any other identity. This is the reverse side of the coin from taking on the professional identity. Retaining a sense of the *non-teacher* identity is, for many professional teachers, also a vital part of the role.

Our final word on developing the 'act' of teaching science is to note that every lesson is a unique experience for you and the students. Even if an experiment or activity may be repeated, the circumstances, people, situation and other factors will be different. This means, to us, that each lesson deserves the best that an inductee or experienced teacher can offer, given the circumstances surrounding the lesson. The uniqueness of each learning experience means that lessons can never really be replicated. This is part of what makes working as a science teacher such a special experience.

THINKING ABOUT PRACTICE

■ In what ways were the teachers in Box 8.1 'acting' to teach science?

■ The incidents in Box 8.1 are described from a student's perspective. How do you think the teachers thought or felt about what happened?

■ How would you prepare to teach a group of students for the first time? List any key points you would wish to make to the students, explaining alongside why you think these are necessary. Compare your list with a colleague and ask an experienced teacher for their advice on what you have prepared.

Chapter 9

Science teachers' pedagogic content knowledge

Jack: I know nothing, Lady Bracknell.

Lady Bracknell: I am pleased to hear it. I do not approve of anything that tampers with natural ignorance. Ignorance is like a delicate, exotic fruit; touch it and the bloom is gone. The whole theory of modern education is radically unsound. Fortunately in England, at any rate, education produces no effect whatsoever.

(*The Importance of Being Earnest*, Oscar Wilde)

INTRODUCTION

In this chapter we review aspects of the expertise science teachers need in order to make learning happen in their lessons, such that students do not find themselves knowing nothing about science. Teaching does not just require teachers to know their subject – but also how to put it across in order that non-expert learners can make sense of the subject and construct their knowledge and understanding. The strategies and techniques used by the teacher to do this constitute 'pedagogical content knowledge' (PCK). Learning the SCIENCE behind the NC is one thing – but learning to teach it is quite another. The purpose of teaching practice is to give trainees opportunities to develop PCK in supportive environments. Part of the process of becoming a professional science teacher is to experiment, not just by doing practical work, but to work on explaining science, giving students active learning opportunities, reviewing their learning and assessing progress. Testing one's own ideas, stretching creative 'muscles' and starting to develop a sense of being comfortable working as a professional science teacher are all part of the process.

We need to say something about how the management of some schools approach having a trainee working in their science departments. We are very critical of school mentors in science departments that are so focused on preserving their own ways of teaching that trainees are prevented from experiencing this experimental phase as they should – with freedom. We have found mentors working under pressure from Heads of Department or senior school staff, who insist that trainees adopt the strategies laid out in the lesson plans already prepared in the science department, under the guise of needing to 'safeguard students' results'. This policy, we believe, creates frustration, stifling professional development and creativity, and is not the purpose of teaching practice.

PCK and its development is, like many issues in this book, a big topic. We offer readers a summary of PCK that have been found useful in our experiences and those of other science educators. Our expectation and hope is that trainees and inductees will take the challenge that developing PCK offers, developing their skills as best they can in order to become true professionals.

PROBING PRIOR KNOWLEDGE

Finding out what your students know already about a science topic is an excellent way of beginning to teach them. Armed with this information, a teacher is in a strong position to know what to do next, what needs further explanation or clarification and how best to differentiate in order to move students' thinking to a more orthodox or accurate scientific view. The Australian science educators Richard White's and Richard Gunstone's excellent text entitled *Probing Understanding* (1992) describes many techniques in full detail, together with other strategies. We offer here a selection both from the text and our own experience that may suit inductees.

Probing students' prior knowledge can be done either at the start of a topic or a lesson, depending on what you are teaching. The time required may be as little as five or ten minutes, or as much as 30 minutes, depending on the strategy chosen. Here are some examples.

Probing prior knowledge about a topic

This may require a longer period of time, such as 30 minutes. Strategies to try are:

1 Developing a concept map (see also pp. 139–40). Students are given an A3 piece of paper and a set of cards/Post-it® notes each bearing a word representing a concept. A corner of the paper is marked off as 'Park'. At the start of a topic, students are invited to place as many concept words as they think they know into the main part of the sheet, making pencil connections between them. Along the joining lines they may write a phrase or word explaining the link. Any concepts they do not know or understand must be placed in 'Park'. At the start of a topic, students may not understand many of the words, so quite a few may end up in the Park. Also, connections may be faulty in some way. As lessons in the topic proceed, students are asked to review their maps regularly, giving them the opportunity to move concepts to and from Park, and to review the connections. The aim is to develop a more accurate concept map at the end of the topic than they had at the beginning. This technique is useful as students can see their own progress. The teacher must be focused very carefully on including all the concept words explicitly in the lessons, inviting students to think about how these relate to each other, but leaving students room to think for themselves.

2 Using 'discussion boards'. This technique can be used in other settings besides exploring prior knowledge. The technique requires a set of laminated A3-sized sheets of card and write-on, wipe-off dry marker pens in different colours. The author's set of cards is made in luminescent colours and was prepared at a commercial photocopy shop. Check that the marker pen ink does wipe off fairly easily. Divide the students into groups, preferably of three or four. For a group of 25, this means you will need about seven boards. Ask each group to appoint a spokesperson to explain the beliefs of the group to the whole class and someone to write on the discussion board. Pose at most three key questions about the topic to the students. Ask them to each consider their own individual views first, perhaps writing these down, then to discuss in their groups so they reach consensus. The consensus views are written on the boards, and then explained to the class. The discussions will probe students' thinking, making them more aware of different views. The short presentations will alert the teacher to the range of views present. Students can record their own views in their notebooks. As the topic proceeds, students can revisit the groups and boards, looking at whether their views have changed. At the end of the topic, a final session will reveal what students have learned.

Probing prior knowledge at the start of a lesson

These techniques require five to ten minutes.

1 *Predict, observe, explain (POE)*. This task requires students to do three things – to predict the outcome of an event, to observe what actually happens and to explain any differences between the prediction and the observation. For example, many students find conservation of mass in a precipitation reaction difficult to understand, thinking that the mass will increase overall because a solid is produced which is more dense than, as they see it, the two liquids which reacted together. A POE task taking ten minutes will reveal the extent of this alternative conception among the students. To do this, a teacher will need to demonstrate a precipitation reaction using two vessels each containing a liquid which, on combining with the other, will produce a precipitate. The teacher should start by showing what happens when the two liquids mix, by demonstrating using small quantities in test tubes. Then move on to the balance. Here, two vessels containing the same liquids are placed on the weighing area. Students should record the total mass before the reaction and *predict* the total mass after the reaction. Then the teacher pours one solution into the other, replacing the empty vessel on the balance. Students *observe* that the mass is unchanged. They then have to *explain* any difference between their prediction and observation.

2 *Making a drawing*. This task requires students to draw their view of a scientific event or concept. For example, when teaching about air to a group of 11-year-olds, one of the authors invited the students to each draw their image of how the air looked 'if they could see it through a very powerful microscope'. A number of different images was produced, expressing a range of ideas but, of course, only a very few showed tiny particles. The lesson aimed to encourage students to change their thinking by offering experiments with air, the results of which could logically be explained by adopting a particle view of matter. At the end of the lesson, students were asked to make a second picture showing how their thinking had changed.

3 *Prompt questions*. This technique is similar to 'Discussion boards' but does not take so long. Prepare a sheet with three or four key questions related to the learning objectives for the lesson and/or alternative conceptions you suspect the students may have. Under the questions, divide the

sheet into two columns, one headed 'Start of lesson' and the other headed 'End of lesson'. Hand a copy of the sheet to each of the students as they enter. They can then focus on answering the questions at the start of the lesson, writing their answers in the spaces provided. The answers may or may not be reviewed as a class. At the end of the lesson, the sheets are taken out a second time and the answers rewritten. Finally, the teacher collects the sheets for formative assessment. One author saw this technique used very successfully by a trainee on teaching practice. The important point was to ensure that the lesson plan allowed enough time to get back to the sheet at the end.

TEACHING AND LEARNING STRATEGIES

We have stated earlier, but it bears repeating that the key point of a science teacher's act is to enable the students to construct their science knowledge. We believe strongly in the principle that teaching and learning strategies must be active rather than passive, requiring students to interact cognitively with the tasks they are asked to do. We do not support lengthy sessions of copying texts from boards, screens, books or whatever; use of meaningless word-searches which are no more than time fillers; making students watch a video without any support material; students carrying out trivial experiments; or teachers lecturing excessively. What we do envisage in an effective science lesson is a series of inter-related 'episodes' that engage and challenge students appropriately.

The starting point for selecting appropriate strategies must be the learning objectives the teacher wants to be met in the lesson. One way of expressing these meaningfully is to put them as 'Key questions'. The strategies chosen will then aim to provide opportunities to answer the key questions, so that by the end of the lesson students may know, or be working towards, these.

Having set the key questions and/or learning objectives, the next task is to decide on activities to enable students to learn what is expected. This can be approached in terms of episodes lasting 15–40 minutes, depending on the length of the lesson and the type of activity. A list of strategies is, in no particular order:

- student-led activities, e.g. discussions, debates, argumentation activities, peer explanations, 'buzz' groups;
- role-plays;

- students making oral presentations;
- students making visual presentations;
- research from paper or ICT resources;
- research/investigation through practical experiment;
- teacher-led question and answer, explanations, demonstrations;
- student experiments/practical work;
- text-based activities, e.g. Directed Activities Related to Texts (DARTs – see pp. 146–7), card 'games', using newspaper articles, writing frames;
- making concept maps;
- 'artistic' activities, e.g. writing poems, students making a video, devising their own role-plays, posters and information leaflets.

Two other valuable strategies that take longer are:

- visiting a science-based workplace or industry;
- working with a guest scientist.

This list is not exhaustive, but comprises activities the authors have either used or seen used in school science lessons. Good practice would be to combine two or more episodes using strategies from the main list in a lesson. Our experience is that lessons are typically one hour or one hour plus a few minutes more. Working on the basis of one hour, if about ten minutes are allowed to get students settled and into the lesson and deal with any administration (register, returning marked books etc.) and ten minutes at the end to round up, there are about 40 minutes for the main part of the lesson. Students' attention spans can be short, hence our recommendation to have two or perhaps three episodes, depending on length and complexity.

The list may appear surprising to inductees used to science lessons comprising mainly lectures ('teacher talk') and practical experiments. In fact, practical work is thought by many science educators to be of limited value in aiding students' learning (see pp. 234–5), so use of many of the other strategies may be more beneficial. Certainly, bearing in mind the variation in learning styles and abilities among a group of students, there is a need to ensure variety in the range of activities presented. Not only does this help to make the subject come alive, but maintains stimulation, interest and challenge for the students, as well as the teacher.

Texts are available which work through teaching strategies in detail. Among those we recommend are *Teaching Secondary Science* (Ross *et al.*, 2000) and *Language and Literacy in Science Education* (Wellington and Osborne, 2001). These two texts give explicit examples of activities teachers may use to aid students' learning in science. We discuss some strategies listed above and from these texts in more detail.

Text-based activities

Science is full of vocabulary that is not easy for students to learn. Traditionally, science textbooks have high reading ages, which can be off-putting for many students. More recently, textbooks have tried to take this into account by presenting material in a more 'user-friendly' format. This is often a 'double-page spread' presenting a topic in only two pages, using short sentences and lots of colour pictures and diagrams. Each two-page spread is designed to make one lesson of material. The danger of this approach is that the science has been reduced to a very low level, making it appear like a glossy magazine rather than a textbook.

However, science teachers need to help students learn the language of science. A range of strategies can be used to help with this.

Directed activities related to text (DARTs – see pp. 146–7) encourage students to focus on text, perhaps stimulating discussion or considering their interpretations and reflections. DARTs can involve text reconstruction – that is, filling in missing words, phrases or labels or re-ordering a muddled list. Other DARTs are more analytical, for example, asking students to find target words and/or phrases in a text, then to work out their meanings; or to convert the text to a flowchart, table or diagram. The analysis can be done as a group or individually. DARTs can be purchased commercially, but teachers can also make their own from text materials such as textbooks, science magazines or short sections of popular science books.

Newspaper articles and other sources of science of current interest can be used in various ways. For example, to stimulate discussion by setting questions for response using discussion boards; to stimulate critical thinking about a science issue; to raise awareness of an issue; to introduce vocabulary; or to prompt production of a leaflet, role-play or poster. Most of the quality UK papers carry a weekly science page often with very good diagrams and accurate text. With care, these articles can be made suitable for KS4 students. Tabloid newspapers should also be used, as these are read by a majority of adults and students. The text level is normally more readable and their dramatic approach provides a good vehicle for introducing critical thinking.

Writing frames are useful to help students approach writing scientifically. The traditional structure, 'aim, apparatus, method, results, conclusion', is a frame we often see used to support 'writing up' of practical work. Without spending time analysing the validity of this, we wish to introduce others that can be used to support other types of writing. Two examples are given in Box 9.1. Frames give students a lead into what you are expecting them to write. Depending on the students' writing skills, these can also be differentiated readily and fairly discreetly.

BOX 9.1 WRITING FRAMES

A frame for report writing	A frame for explanations
Title	Title
The animal I am describing is . . .	I want to explain why . . .
It normally lives . . .	An important reason for why this happens is . . .
It feeds on . . .	The next reason is that . . .
During the day/night it can be seen doing . . .	Another reason is that . . .
It is able to survive by . . .	

(Wellington and Osborne, 2001: 69 and 71)

Card 'games' are an excellent way of introducing scientific vocabulary. One of the authors has often made sets of 'formula cards' to help students learn to write formulae and equations (Barker *et al.*, 2002). These were used to provide a structure so that when students had memorised the information after several weeks of use, the cards could be removed, as the students had gained the necessary confidence to write formulae and equations without them. Other card 'games' can be used to help introduce related items, for example, organisms in a food web, parts of an electrical circuit or classification exercises. The kinds of tasks that can be presented include: sorting correct, incorrect and partly correct answers to one or more questions, matching pairs, matching questions and answers, ordering instructions, ordering to make a flowchart, constructing a diagram, matching pictures and titles. Each group or pair of students will need a card set.

Our experience suggests that students enjoy opening an envelope to take out a nicely prepared set of cards – it is rather like receiving a present, as there is an air of mystery about what they will find. Laminating the cards makes them more durable. It may seem that this type of activity is suited best for younger or weakly motivated students, as card games will not seem like real work. However, in many schools, card sorting activities are very common in humanities teaching, and students recognise when an activity has some genuine 'content' as well as being fun. One of the authors developed a rather involved card game to help A level students who were having difficulty grasping the concept of 'core charge'. Students soon spot activities that are not stretching them, and seem designed only to keep them occupied. They do appreciate imaginative ways of making learning interesting and different.

Student-led activities

Teaching strategies allowing students to talk actively, to take the lead in explanations, discussions and arguments about science issues and concepts can be very fruitful in terms of achieving learning outcomes. Teachers may be surprised at how involved students become in such activities and how seriously they take them. The purpose is to give students the chance to develop and test their ideas about scientific concepts and topics. These activities are similar to the kinds of ways scientists discuss ideas.

Student-led discussion, arguments and debates need to be carefully structured around specific items. We also advise using discussion boards or posters to help provide feedback.

Argument Students are presented with a short text outlining an event involving a scientific topic or concept, for example, an account of an experiment. We can use the precipitation reaction (see Predict, observe, explain, p. 182) as an example. The account of the experiment is written out on a card, perhaps including a diagram. Underneath are three choices, such as, in this case: the mass shown on the balance will be higher after the reaction; the mass will be the same; the mass will be lower after the reaction. Students must first decide what they think and then discuss their answer with group members, working to reach consensus. This means that they must be able to express a reason for their answer. A second way of using argument is to ask students to discuss reasons why an event takes place. For example, asking a question such as 'What evidence is there that matter is made from tiny particles?' may be followed with several different possible reasons. Students have to select the reason they most closely agree with, then in groups argue for their reason. Rather than having to reach consensus each time, groups could also present each statement together with the arguments offered in support. Wellington and Osborne (2001) offer further suggestions.

Discussion and debate When structured carefully, students show great willingness, enthusiasm and seriousness over discussions. The authors have found that rather like role-play exercises, these often require significant preparation beforehand, to ensure that the topic is clear and that there are rules everyone can agree with, but that once running, discussion activities are excellent ways of encouraging learning. Care should be taken in organising the groups – the ideal group size would be about four. Too few and there is insufficient discussion. Too many and the group becomes difficult to control. A set of rules is essential. The task must be clear, that is there must be a definite outcome that students are expected to achieve. Discussion boards or posters to summarise the key points are essential.

Peer explanations, 'buzz' groups A good way of summarising a question and answer session, or teacher-led discussion or explanation, is to ask students to explain to someone else in one minute the key points about what has just been discussed. The receivers of the explanation then have a minute to indicate if they agree or disagree with the explanation. If this process is repeated during a discussion, then students have the opportunity to swap roles. This activity has the advantage of being short, breaking up what may be a lengthy session of concentrated listening to teacher talk, and helpful, as the teacher is able to listen in on what is being said. A way of extending this is to invite explainers to talk to groups of students, the whole class, or younger students. This latter activity requires much more preparation on the part of the student(s), as they need to be confident and accurate in their material. However, the time spent is repaid in the satisfaction obtained.

We will leave the discussion of strategies in the hope that we have stimulated some creative thinking about the kinds of activities inductees may try in their lessons. We strongly encourage experimentation and use of creativity to go beyond the 'normal' range of practical work and explanations. Be prepared to take risks on trying new strategies, preparing imaginative activities designed around stimulating, relevant material and letting the students have the freedom to test their own ideas out. By doing this you will become an exciting teacher in whom the students can have confidence.

EXPLAINING AND MAKING MEANING

A critical factor in teaching science is being able to explain difficult ideas clearly, such that non-scientists can understand what you mean. Significant research has been devoted to exploring ways in which teachers explain science ideas. Two texts discussing the issue of explaining and 'making meaning' in school science are *Explaining Science in the Classroom* (Ogborn *et al.*, 1996) and *Making Meaning in Secondary Science Classrooms* (Mortimer and Scott, 2003). Rather than review these extensively, as we cannot do them full justice in the space available, we focus here on the kinds of explanatory strategies science teachers adopt.

Explaining science through dialogue

Discussing a topic with students is perhaps the most obvious strategy for teachers to use. Dialogue comes in many different forms. For example, at the start of a lesson, a teacher may use dialogue to *open up differences* (Ogborn *et al.*, 1996) between what students do not know and what they need to know, or between what students think they know and what the teacher knows. This creates expectations among the students for what is to follow. A second type of dialogue involves *construction of entities*, in which teachers help students develop knowledge, understanding and give meaning to scientific vocabulary. Here is an example of a teacher introducing density to a year 9 class:

> *Tom*: Right, listen, what we want to do now is have a chat about ... what this word means, density.
> *Student*: I know.
> *Tom*: Now, I'm going to tell you what density means, and then we're going to go and have a look at how you can measure it.

189

Student: Is the same as on a disk?
Tom: No, well, in a way, what you mean a computer disk?
Student: Yeah.
Tom: How much information you've got? In a way it is.

(Ogborn *et al.*, 1996: 42)

The students are being introduced to something new, and as Tom, the teacher, introduces something to them, so one responds with his/her own understanding of the scientific term. This is also an example of how an analogy can arise almost of its own accord. The student is familiar with computer disks, so has heard the term before. The teacher, though, has to move the lesson on towards his meaning for density, so engages them in activities involving wood and metal so that density becomes related to *things* rather than being just a *word*. Helping the students construct a scientific meaning for density for themselves presents quite a challenge and may not be solved in one lesson alone. Rather, students will need to assimilate the information from this lesson and take it further before the concept of density is correctly constructed.

Question and answer sessions with students can also help build up their knowledge and understanding, helping to make meanings for scientific ideas. This can begin by asking students what they think about an event, topic, concept or issue. In this approach, it is vital to accept the responses given, however incorrect they may seem at first. The process of questioning can help to sort responses out, taking care to ensure that students are not belittled. Once students' initial ideas have been obtained, questioning can continue towards building scientific knowledge. Here is the first part of an example from Mortimer and Scott (2003: 51–66), in which students placed nails in different places to find out the conditions that would create rust. The teacher reviews where the students left the nails and then reviews what it was about the places that made the nails go rusty. Part of the dialogue was:

Teacher: So – what I want to do – put on the board, is perhaps put down your ideas of what it was about the places that made your nail go rusty. What do you think it was – thinking about the places – that made your nail go rusty?
Haley: Damp.

Teacher: Damp. Now, we'll put things up first of all, then we'll have a think about them in a minute. Right, so damp [Teacher writes it on the board]. Yes, Cheryl?

Cheryl: Moisture.

Teacher: Moisture [writes it on the board]. Damp, moisture. Anything else? Gavin?

Gavin: I put mine in some mud in the garden.

Teacher: What was it about the mud that you think made yours go rusty?

Gavin: Cos it were all wet and all boggy.

Teacher: Wet – so it was wet again. Wet [writes it on the board]. Right – any other ideas, Matthew?

Matthew: Air. . . .

(Mortimer and Scott, 2003: 51)

The teacher continued like this building up the list of words 'rain, damp, moisture, wet, salt, vinegar, air, condensation, cold, dark' (p. 53). After more discussion, the teacher helped the students see that several of the words meant the same, so the final list became 'water, salt, vinegar, air, cold, dark, dry' (p. 55). The teacher's next stage was to move to an activity in which the students constructed a grid matching the characteristics listed against the places where the students placed the nails. The idea here was to try to identify from the list the *essential* conditions that made the nails rust. The sequence of discussion and activities continued until the students' thinking was consistent with the scientific viewpoint.

Time and experience are required to develop this skill. Inductees should be aware of what they are aiming for, working to develop their own skills in questioning and leading students. Experienced teachers have the ability to make this look very easy. The temptation at first is to jump in too soon with the 'correct' answer, without ensuring that this has meaning for the students. Patient work with students' ideas matched with sensitive questioning will gradually lead them to the desired point.

Using analogies

Everyone uses analogies in their speech and teachers are no exception. Analogies involve using phrases such as 'It's like a . . .' and 'It's as if . . .'. In teaching science, analogies provide useful ways of helping students move forward in their learning, provided these are not taken too far.

191

Experienced teachers become used to using analogies in their explanations, but it takes time to develop these into everyday practice. For example, one of the authors taught atomic structure in A level chemistry for several years. To help students understand the notion that energy 'comes in packets called quanta', she bought a multi-pack of small packets of crisps. The purpose was to illustrate that crisps, like energy quanta, only come in packets, unlike sweets, that can be bought loose. We did not take the analogy much further, but it did make the point that the amount of energy in a quantum is fixed and cannot be changed, and that just as crisps have different flavours, so quanta have different amounts of energy.

Making models

Modelling a real-life situation is a common strategy in SCIENCE, as often the ideas or events are impossible to 'see'. Using models in science teaching is therefore an entirely valid way of helping students to 'see' what cannot normally be seen. The models may take various forms. For example, in helping students to understand the structure of DNA, building models using any of a variety of materials, including card, Lego™, sweets, straws or paper (see *School Science Review* March 2003, 84 [308] for examples and ideas) is a good strategy for helping students to appreciate the complexity and size of the molecule. Chemical model sets are used widely to help students understand other molecular structures and chemical equations. Other examples of modelling include asking students to construct their own models based on their ideas about a specific concept, by making a drawing (see p. 182). The lesson can proceed by looking at differences between the models and comparing these to experimental evidence. A third possibility is to present an abstract idea using a model. Two different types of beans in a jar can be used to represent molecules of two different liquids that may mix on shaking. Commercial models of specific items are also widely available – schools may have models of the body, or body parts and a vibrating particle model to illustrate changes of state. Any type of model can act as a prompt for a discussion with students about their ideas, helping them to construct scientific knowledge.

Telling stories

Story-telling is helpful to capture students' attention about an issue or topic. Students normally appreciate a good story, especially if it is true.

 192

Touching emotions can bring about surprisingly effective learning environments. Some teachers fall naturally into a 'story-telling mode', while for others this strategy is a comparatively rare event. A good narrative story can carry a good deal of knowledge, as well as being memorable. Ogborn *et al.* (1996) give an example of a teacher who told a long story about the actor John Wayne's stomach cancer as part of a lesson on digestion. Here is an excerpt from the story:

> *Teacher*: Do you all know the film star called John Wayne?
> *Students*: Yeah.
> *Teacher*: Yeah? John Wayne. This is a long story, settle down, it's a long story. John Wayne was making a film in the 1940s and when he was making the film in the 1940s, they were testing atomic bombs in the valley next to the one, they were making a western film, so it would be a film in, you know, the Wild West, and – and, John Wayne was kind of filming with, with all the cameramen and all the other stars . . .
>
> (p. 122)

The teacher goes on to explain that John Wayne had his stomach removed and had 'tubing' joined up inside so he 'could manage without the stomach' (p. 123). The overall message was that 'stomachs digest proteins' and this message came out despite the teacher making detours into other issues. Although the story is rather gruesome, it does contain science knowledge, including information about how people might get cancer, how this may be treated and what functions parts of the digestive system have. Other examples of stories can be about scientific discoveries, the lives of scientists, news events or personal stories.

An alternative strategy for story-telling is to involve students in the process. One of the authors was observing a lesson in a laboratory and noticed a display focusing on 12–13-year-old students telling stories about their own births, as part of a series of health education lessons. The teacher (not the one being observed) had, as part of the activity, asked the students to find out what happened when they were born, each bringing in a picture of themselves as a newborn baby and writing about what happened. The stories made a fascinating display that was both emotional and scientific. An issue like this needs sensitive handling, but with care could make a valuable contribution to students' understanding of sex, child development and health.

When undertaking discussion and dialogue with students, use of praise and making good eye contact is a vital part of the transactions. Students need to feel respected and valued for their contributions, both by the teacher and other students. Part of the teacher's role is to create a good atmosphere for sharing information and ideas. This requires good classroom management and a sense of mutual respect. Neither of these comes easily to most trainees, but need to be worked on with time. Beginning teaching with a strong, positive approach emphasising and valuing individuals but also setting rules for behaviour is a good starting point.

ENDINGS AND OUTCOMES

Here we review, briefly, endings for activities and lessons and outcomes. At first, inductees must expect that finding time to end activities and lessons neatly can be difficult, as judging how long it takes for students to do what has been planned is not easy. We have seen many lessons where activities take too long, leaving no time at all for review as well as lessons where the trainee has been left with 10 or 15 minutes to fill. In the newness of learning how to teach, reviewing learning tends to be pushed aside. Even experienced teachers do not always review learning in meaningful ways, so inductees do not always see good practice. We see reviewing learning as an important part of a lesson and something all inductees should work towards as part of sound professional practice.

Most activities carried out with students need to be reviewed afterwards. This might mean going through answers to questions; leading a question and answer session; asking what students' ideas are having completed an experiment or other activity; or analysis of written work. The teacher's task following this is to make a transition to the next activity, if this happens in the middle of a lesson, or to review the learning objectives if at the end of a lesson. Reviewing learning is part of formative assessment (see Chapter 10), so gaining an understanding of where the students have got to is necessary for working out what to do next.

In reality, most lessons will result in a variety of outcomes for the students. Some will have connected with the material as envisaged by the teacher, while others will still be very distant. Most will probably have engaged with something, but may still be muddled in some ways. The task of developing students' knowledge requires patience and time. Here are two extremes to be prepared for.

'Aha!' moments and the 'Eureka!' feeling When students have understood an idea which is new to them in the way the teacher was expecting, this can be extremely exciting. Evidence for this can come out in any aspect of a lesson – and is especially rewarding. One of the authors calls these 'Aha!' moments, as they can happen quite suddenly (cf. Koestler, 1979: 110). A student may just 'connect' with the new idea and become really excited as if s/he has made a great discovery – which of course, for him/her this will be. Often, this situation is infectious, as the student who has 'understood' will then be pleased to share what s/he has found out with others. Make the most of these moments, working towards developing strategies for ensuring they happen.

Trying again when it all goes wrong The other extreme is that no one understands what was expected and the lesson ends with a sense of failure. This can be hard to take, especially if a trainee thinks that they prepared well with good activities. Often, the reason everything 'went wrong' is more a classroom management, that is, discipline, issue rather than a fault with the scientific material. As we described above (see p. 174) this happens to everyone at some point. The main way to deal with the feeling is to keep a sense of perspective – and, like falling off a bicycle, to get back in to the laboratory and try again, if necessary with new activities, a clearer, more accessible explanation, analogy or model, a more emphatic discipline system (with back-up if possible) and determination to succeed.

THE IMPORTANCE OF BEING YOURSELF

Teachers are unique individuals, just as the students are. As a trainee, even if you have excellent mentoring and, perhaps, especially if you feel forced to conform to an in-school style of working, this is something you must always bear in mind – that each teacher has to be him/herself, not a clone of anyone else. Of course, you may pick up tricks, activities, analogies, explanations, key questions and other tips from colleagues, but in the end you have to develop your own professional act, which really is yours and no one else's. Your professional teacher identity will reflect your own character, preferences and style that students will come to respect. With time, if teaching is the job for you, you will feel comfortable and be able to be yourself in front of students. This will not happen immediately, but will be a gradual process perhaps taking several years. In this way

195

teaching is a marvellous job, because you truly do have the fantastic potential to, in a way, 'run your own business' creating the best possible opportunities for students to work with you in your way to learn science. Not only is this a great challenge, but it also brings great rewards.

A second factor we believe leads to the sense of achievement and reward is the ability to work outside one's 'comfort zone'. We have continually been surprised at how quickly inductees settle into a predictable pattern of classroom activity, perhaps based on expectations placed on them by the schools in which they work, on the role played by an inductee's mentors in assisting their development or simply by their own choice. Although an element of certainty about one's own practice is necessary, getting into too fixed routines and practices leads eventually to a sense of sterility and possibly boredom. If the teacher is bored then the outcomes for the students are not positive. We strongly encourage inductees to look for new activities or strategies to try, to make up their own, to keep variety coming in their lessons. Doing something new for the first time can be time-consuming, scary and uncomfortable, but it may well bring unexpected rewards. This is what we mean by working outside the 'comfort zone' – it is too easy to stay within it.

Being yourself will, in the end, bring the best out of you and help you give the best to your students. It is extremely rewarding when students show their appreciation of what you do, or appreciate you for being yourself.

THINKING ABOUT PRACTICE

- Choose a topic from the NC Science with which you are familiar. Use information from this chapter to devise: a key question you want students to be able to answer; a way of probing students' prior knowledge; an activity to help students learn; and questions to review learning. Test your results on a colleague or critical friend and discuss any learning outcomes.
- How would you approach working as a trainee in a school where you find yourself 'required' to do things in the school's way?
- Observe a science lesson being taught in a secondary school. What activities were undertaken to help students learn the topic being taught?

Chapter 10

Evaluating teaching and learning

Il est plus nécessaire d'etudier les hommes que les livres.
(It is necessary to study people rather than books.)

(La Rochefoucauld)

INTRODUCTION

This chapter is about evaluating teaching and learning. At the end of the chapter we will focus on evaluating teaching explicitly. We start with a consideration of assessing *learning*. It should be clear to any reader that if teaching and learning are intimately associated, then evaluating teaching must, in part at least, be informed by the assessment of student learning. 'Teaching' that does *not* facilitate learning, after all, can hardly be considered successful.

One of the points we made in Chapter 7 was that professional teachers should have a rationale for their actions. Our work is resourced mainly through public funds, and we are entrusted with the education of young people and given authority over them (*in loco parentis* for those under 16). These are strong reasons why we should be able to justify what we do. Accordingly, there should be reasons for asking a group of students to undertake a homework; why we use up consumables in a practical; why we use 30 minutes of lesson time viewing a video, copying notes or requiring students to copy from the board. If we cannot justify an activity in terms of meeting specific learning objectives, this may not be good use of time and other valuable resources. In practice, experienced teachers do many things out of habit and, being very busy, we may readily fall into such a trap. For example, running a favourite practical activity can be tempting, even if it is no longer relevant to the latest version of the NC.

In the real world, constantly asking 'why am I doing this?' before *every* professional action is not viable. The work of a teacher means that a great deal of our work involves making decisions at a tacit level, 'on autopilot'. Nonetheless, regularly reviewing the way we do things to make sure we still agree with the original rationale ('is this *still* the best way to do this?') and that the rationale has not become redundant ('have our objectives changed here?') is essential.

Similar points may be made about assessment and evaluation activities. Assessing student learning is central to the teaching process, as is evaluating our teaching, but we should make sure that the means we use to assess and evaluate justify the time, effort and potential for causing stress.

SUMMATIVE ASSESSMENT

External examinations

We assess for different purposes, so need different assessment tools. The type of assessment that has the highest profile in the public eye is that of formal examinations such as GCSEs and A levels. More recently SATs taken at ages 7, 11 and 14 have also become high profile as they are used to judge schools' performances publicly. These are examples of *summative* assessment, when the outcome is a grade or mark indicating the general standard of performance at the end of a period of study. GCSE grades are meant to offer a summary of learning by age 16, at the end of compulsory schooling. Summative assessments are used commonly for selection; either for admission to the next level of education or for a job.

Summative assessment may be norm- or criterion-referenced. Norm-referencing gives an outcome in terms of the cohort. For example, the students with the top 10 per cent of marks in an examination may be awarded 'merits', regardless of their actual scores. Criterion-referencing awards grades according to whether students have met set criteria. 'Merits' may then be awarded to those meeting all the criteria. The NC level descriptors are criterion-referenced; if every student meets the descriptors for level 8, then they may all be assigned this level. Gaining QTS is criterion-referenced; any trainee demonstrating they have met all the criteria should be awarded the qualification.

Specifications for external examinations such as GCSE and A levels often include specimen grade level descriptors. However, as the examination itself consists of large numbers of examination questions covering different skills and areas of knowledge, finding a simple relationship between students' scripts and the grade descriptors may be difficult. The Sc1 criteria, for example, are designed to be hierarchical, in the sense that a student should not be considered for 8 marks unless they have met all the criteria for 6 marks, even if they show evidence of being able to meet the 'extra' more difficult aspect of the criteria that would otherwise allow 8 marks to be awarded.

Using this type of criterion-referencing in complex examinations can be very dubious, as a very able candidate may fail to demonstrate one of the criteria required for a low level score. Imagine if Einstein submitted a paper on relativity, but failed to 'give all answers to an appropriate level of precision', so was limited to a D grade no matter how creative and accomplished the rest of his work was.

In practice, in external examinations, the grade awarded relates to the sum total of the marks scored (once any adjustments are made for relative weightings between papers), so a candidate scoring 126/180 would have to be awarded at least as high a grade as a candidate scoring 125/180. This would not always be the case if pure criterion-referencing was used, as when we award young Einstein a D for his 178/180, but give a B to another candidate who scores only 148. External examinations such as GCSE have evolved into a curious mix of criterion- and norm-referencing, leading to regular media debates on whether 'grade inflation' means standards are improving or being lowered – in both cases because more students are making the grade. We should recognise that public (*sic*) examinations are complex, influenced by political and commercial agendas.

The political philosophy of some parties and pressure groups relates strongly to notions of providing equal opportunities for all, with a belief that people generally have high potential that can be blocked by socio-political factors and institutions. These groups may see movement towards a society where half the population hold university level qualifications as a sign of progress. Others hold different philosophies, believing that societies should have elites that are recognised and supported. This need not be in terms of aristocracy and birthright privilege. A meritocracy is also possible, such that the most able, who can potentially (it may be argued) contribute disproportionately more to society, are able to demonstrate their quality. In such a society, a competitive education system where the elite can thrive but the majority are filtered out is more beneficial. Our intention is not to offer a critique of political theory or oversimplify the positions of political parties, but we emphasise that decisions about how to grade examinations are informed by political stances and are not purely educational decisions.

Mark schemes and examiners' reports are now widely available to teachers, who benefit from this added insight. These documents provide important hints into the ways examiners conceptualise the examination specifications, and where marks are commonly awarded. This is important for teachers, as school science is not a 'given' that automatically follows from considering how to teach aspects of science. There are many possible ways of transforming knowledge from SCIENCE to create school science. In practice, curriculum science is an entity socially constructed through the interaction of curricular specifications, teaching and the formal assessment system. School learning is only formally recognised when it leads to answers considered correct in examinations, so teaching may become most valued when it helps students give those answers in examinations.

Summative examination grades are compiled across the UK to produce 'league tables' that are published annually. Parents use these to help make decisions about which schools to send their children to, while schools use them to publicise achievements or improvements. Hence, there is a strong tendency for some teachers to 'teach to the examination' to give their students the best chance of scoring high grades. Other teachers recognise that examinations do not reflect the full range of objectives for science education, believing that this approach 'short-changes' students. We suggest that keeping students interested and motivated through exciting and varied lessons is ultimately a more effective strategy for achieving good examination grades. There is undoubtedly a major professional issue for teachers today: at least for those of us who believe that the examination system inevitably distorts the aims of education.

Writing examination questions

Our intention is not to be critical of the examination boards. Setting fair and valid examinations is a complex and technical process, undertaken by examination boards on the basis of considerable research and development in producing their papers and procedures. The boards, like teachers, are constrained by the QCA's external requirements so their scope for innovation is limited. For example, a trend in recent years has been to help candidates by setting questions in more familiar contexts. This is well meant, but ultimately has problems. Context-based questions can reduce the validity of the examination, because what is application of a principle to one student may be recall for another, making the question 'unfair'. Using contexts can also make it more difficult for some students to understand what is required (Taber, 2003f). We suggest that writing examinations is difficult and we advise teachers to seek specific training if they intend to write examination papers for use in school.

DIAGNOSTIC ASSESSMENT AND ASSESSING PRIOR LEARNING

Diagnostic assessment is another form of assessment valuable in teaching. Specialists use this kind of assessment to identify students with specific learning difficulties such as dyslexia. In science teaching, diagnostic assessment is used to elicit learners' alternative conceptions (see pp. 102–7). Assessing prior learning before teaching is as necessary as assessing student achievement *after* teaching. As part of preparing to teach school science you will have identified the essential prerequisite knowledge needed as the foundations of new learning and identified common alternative conceptions that may interfere with the learning process (see Chapter 7). Before teaching a topic, check that the prerequisite knowledge is in place, and find out if students hold ideas that need challenging during the teaching process.

FORMATIVE ASSESSMENT

Assessment for learning

In recent years there has been a major shift in thinking about assessment, from focusing largely on assessment of learning at the end of a course to what is known as assessment for learning (Black and Wiliam, 1998). Formative assessment is used for informing teaching and learning, rather than just judging outcomes. To some extent there is overlap here with diagnostic assessment, which also informs teaching. Formative and summative assessments differ in terms of *when* they happen. Summative assessment must happen at the end of a course of learning. Formative assessment must happen while learning is in progress, otherwise feeding information back into the teaching–learning process is impossible.

If this were the only difference, then we could argue that something that is summative at one stage of the education system could be considered as formative when viewed at the next stage. For example, GCSE grades inform the courses students can subsequently follow. However, that is not really what we mean by being formative. Information derived from summative assessments concerns an overall level of performance or attainment. Formative assessment provides specific feedback to teacher and student on strengths and weaknesses: things the student is doing well and those needing more work.

The nature of formative feedback

Research shows that merely offering marks and grades is often demotivating for students, as this will not help them know how to improve their work. Offering comments alongside grades and scores may be more helpful in principle, but research also shows that most students focus on the grades or scores, often ignoring the comments. This reflects on the historical importance placed on grades and scores at school. Feedback based on informative comments has the greatest effect in supporting student learning. Praising students for their strengths and providing focused, specific feedback on what they need to do better, has been found to be the best way of improving their work. This message is being widely broadcast in educational circles, but many students, and their parents, still associate *marking* work with *grading*.

In the professional context, decisions about marking policy should be made by teachers. Part of a teacher's professional duty is to base our practice on what research has identified as 'best practice' (see Chapter 12). This task clearly falls within teachers' professional competences. Once again, however, we find that in reality there are still many schools where the expectations of parents (and perhaps some staff with 'traditional' views) hold sway, so teachers are expected to do what parents want rather than what is likely to be best for the students. This is largely a consequence of the way that education in the UK, in common with other public services, has been transformed in the public mind into a service industry where the customer is 'right' and entitled to switch to a different provider if not satisfied. Parents have a major interest in the education of their children, but we do not see that either treating schools as market competitors, or considering parents and students as 'customers' buying a product, is likely to be in the long-term interests of teachers, schools or society. The teaching profession must be free to make decisions based on the best interests of our students. Reducing teachers' roles in the assessment process also reduces the 'professional teacher' to 'professional educational technician'.

In practice, faced with this type of issue, we recommend that teachers work to bring about policy changes at department and school level. If school staff can produce evidence and rationale for policies that may be unpopular with parents, then they are in a strong position to take those policies to the governors to persuade them that the teaching profession knows best. Ultimately, schools where the parents cannot support well-considered professional decisions are unlikely to be viable as successful schools, as they will be unable to recruit and retain the best teachers.

Planning for formative assessment

Formative feedback tasks do not need to *look like* tests. Most work students do could offer useful information about progress. This excludes activities like copying information from books, websites or the board/screen (see p. 146). Indeed, opportunities for formative assessment should be identified as part of the planning process when developing lesson plans. If planning begins from objectives specifying what students need to learn and which inform the lesson activities that best facilitate learning, formative assessment opportunities can be built in quite easily. This means that because formative assessment explores learning *in progress*, separating learning activities from assessment activities is not necessary. The same activities can be used to help students learn and to assess their learning.

STUDENT VOICE, SELF- AND PEER-ASSESSMENT

One current concern in schools is that of 'student voice' (Rudduck and Flutter, 2000). There is recognition that students have had minimal say in many aspects of school that obviously influence their learning experiences. There are various reasons why we *should* listen to the student voice: some are more principled, because we value students and their perceptions, opinions and experiences, while others are more pragmatic, because it may help them engage more with school learning. Certainly, one consideration is the notion that students should have more responsibility for their own learning, as people are more likely to be effective learners if they believe learning is something they do, rather than something done to them (TLRP, 2003).

Part of this drive to acknowledge and tap into the student voice is the use of self-assessment. This means allowing students to have input into judgements about their own progress. This does not mean letting them select GCSE grades or decide they were correct to classify spiders as insects. Rather, as part of the process of on-going assessment and review, students can be involved in target setting and evaluation of their progress. If this helps students take (shared) responsibility for monitoring and directing their studies, then this is likely to be a positive outcome in most cases.

Occasional peer-assessment can also be a useful technique. If we want students to understand criteria for judging their work, then asking them to use those criteria to give feedback to other students can be useful. This is one area where school science can reflect a very important aspect of SCIENCE, where peer-review is so central. Peer-tutoring can also be valuable both because the tutor is more likely to appreciate the level at which the tutee is thinking, and because the tutor has to engage with the material to help their peer. This can be very successful when older students are invited to peer-tutor younger ones.

We should point out that both self- and peer-assessment need to be introduced carefully and in such a way that it is done professionally, and not as an act of professional abdication. The teacher can delegate the activity, but not the professional responsibility.

INFORMAL ASSESSMENT

Assessment does not have to be formal and based on written work. Some assessment, often based on written work, will be formalised and recorded. In these cases, students should be aware that work is to be assessed and told in advance the marking criteria. Even formal assessment may include spoken contributions and contributions to group work, as long as the assessment criteria are suitable. Both oral presentations and teamwork are essential aspects of scientific activity that we would wish to reflect in authentic school science.

Much assessment is not formalised, but involves watching students as they think and work. Much of this occurs through questions. Teachers ask many questions. Some are genuine, in the sense that we do want to know why Sandra has not handed in her homework, or why John has Ali's exercise book. Sometimes, perhaps not often enough, we genuinely wish to know what Louise thinks about wind farms or whether Kamiljit finds circuit work interesting. Many teachers' questions, though, are pseudo-questions. When the science teacher asks, 'Which element has the symbol Sn?' this is not likely to be the question the teacher wants answered; the real question is '*Who knows* which element has the symbol Sn?'.

Much classroom talk takes the form of teacher initiation (often a question), student response and teacher's evaluation of the response (Edwards and Mercer, 1987). We think there should be more open talk in science lessons, with students suggesting and developing ideas. Nonetheless, the initiation–response–evaluation triad allows teachers to build up ideas with students, while monitoring constantly whether they are following the argument. This assessment is diagnostic, formative and informs the highly interactive process of managing students' knowledge construction.

Teacher questions and feedback can also aid the differentiation process in science teaching. Differentiation means meeting the needs of all students, taking into account different abilities and strengths, learning styles and learning difficulties. Each lesson should be planned with differentiation in mind. Various approaches to differentiation can be used. The two main ones are differentiation by task, which means preparing different or modified versions of an activity to meet the needs of varying students and student groups; and differentiation by outcome, which involves providing the same task for everyone, allowing students to interpret this individually, each being challenged and successful on their own terms.

In some subjects, differentiation by outcome is more easily applied than in others. Science provides many opportunities for differentiation by outcome. A writing task asking students to describe the journey of a blood cell passing around the body would be a challenge for everyone. All students could produce something judged a success at their own levels. However, the nature of science, and the volume of school science content in the NC, mean that using differentiation by outcome as a strategy for many activities would result in work that both teacher and student know is not very good. In these situations, modifying the task to fit the students may be more effective.

Differentiation by task has two disadvantages suggesting that it must be used sensitively (see the comment about differentiating writing frames on p. 186), and meaning that it is not as widely used as the inductee might expect. Differentiated tasks can lead to stigma, as obviously some students have been set much easier work. Second, modifying tasks for different groups can become a very time-consuming operation for the teacher.

Good science teachers will differentiate by outcome and by task, but will also use a third approach. This is 'differentiation by support'. Here the same basic task is set for all the group, but the teacher supports learners in different ways and extents, depending upon their needs and the levels at which they are working. This support acts both as a teaching strategy and as a form of on-going monitoring of learning.

Using the initiation–response–evaluation pattern with a group of students requires the teacher to involve *the whole class* in the questioning process. This means that answers should not be called out but the teacher should decide who answers any question. Again, this allows the teacher to make sure that the level of question matches the student, so that everyone can experience some success. As part of the purpose of the questioning is to monitor understanding, there will be many occasions when the students do not know the answer, but at least over time everyone in the chance will get the opportunity to *mostly* reply with answers that can be praised.

EVALUATING SCIENCE LESSONS

Evaluation is an essential part of the professional's mind-set. Any professional has a responsibility to evaluate their work, to ensure it meets the accepted standards and to see if there is room for improvement. Evaluation should occur at school and department level as well as being carried out by individual teachers on their own practice. The starting point is probably in terms of individual lessons.

Each lesson will have a lesson plan developed from the SoW and curriculum documents (see pp. 128 and 142–4). The plan will explain the learning objectives for the lesson, outline the activities designed to achieve the objectives, provide timings and details of the resources (including ICT and any apparatus for practical work) needed. The plan should also indicate how work is differentiated to meet the needs of gifted students and those with special needs, among others. If learning support assistants are to be present, then their role should be specified.

The lesson plan should prove a good starting point for evaluating the lesson. The teacher's evaluation will certainly consider such matters as whether all the intended activities were completed, and whether the planned time was available for them. However, the main criterion for success will be whether the lesson objectives have been met. Although the teacher will have a useful impression of how successful the lesson has been, the evaluation should be based as much as possible on *evidence*. In particular, the teacher is looking for evidence of what students have learnt from the lesson – and this is why teachers monitor student learning continuously during lessons, and by regularly checking written work.

Another useful source of evidence is the students. Occasionally surveying students to find out their level of interest and enjoyment of lessons is becoming more common (see the comments on 'student voice' above). Many teachers also acquire continuous feedback by informally talking to students on a regular basis to gauge their responses to the teaching.

During teacher training and, to a lesser extent, during the NQT year, the inductee will be observed by more experienced colleagues. This provides an additional viewpoint to inform evaluation. The *language* of the teaching standards is such that making judgements is somewhat subjective. An experienced colleague will interpret the standards and relate these to classroom events. We argue that this peer observation could be more common in science teaching. Unfortunately, teachers have so little non-contact time that arranging peer observation is quite difficult. Inviting a trusted colleague to observe and comment on our teaching can be a valuable way of obtaining additional information, support, encouragement and evaluation evidence.

Teachers' work takes place mostly in closed classrooms with the students as the only audience. A professional teacher therefore takes on significant responsibility early in their career, performing mainly behind closed doors. Although teachers generally respond well to this, soon feeling they have autonomy in their teaching room, we do not think this is healthy. Teachers have limited opportunities for feedback on their teaching from those who are in a position to offer informed advice. Preventing this restricts the potential for developing new skills and brings the danger of bad habits becoming fixed in a teacher's behavioural repertoire.

We support teachers having significant professional responsibility and autonomy with their students. However, we also think that there should be more opportunities for peer observations undertaken in an atmosphere of invitation, trust and mutual support. One key aspect of being a professional is accepting a commitment to continuing professional development (CPD), and appropriate peer observations can contribute a great deal in this respect. We will have more to say about CPD in Part III (see pp. 215–16).

THINKING ABOUT PRACTICE

- Select three key points about learning from Table 6.3 on p. 131 to consider when evaluating the success of a lesson. Discuss your choices with other colleagues or inductees. Identify factors that may indicate to you that students have learned as you intended.

- Your school science department's marking policy requires you to give a grade to at least one piece of work from each group of students every week. Your timetable includes a number of groups achieving only low grades, despite the students' efforts. You realise the policy is de-motivating. Do you follow the policy or seek an alternative?

- Choose two activities from Chapter 8 and think how you would formatively assess students' learning from these. What form would your feedback to the students take?

Part III

Reflecting on science learning: the place of educational research

The previous parts explored SCIENCE, school science as a subject, and key aspects of effective school science teaching. This section presents additional issues a trainee might consider in becoming a fully professional practitioner. We make the assumption that a professional should always be a *reflective* practitioner (Pollard, 2002). By this, we mean someone open to new ideas and developments, with the ability to identify professional problems and take action to solve these.

Given the research-based nature of SCIENCE itself, one might expect science teachers to be open to such ideas. In practice, research does not always impact on practice in the ways we might hope and expect (de Jong, 2000). Two major barriers inhibit new science teachers from adopting a reflective attitude to their work. The first reflects the discussion in the first section: few graduates entering teaching directly from undergraduate study have experienced the processes of SCIENCE *directly*, so they lack research experience. As a result, their views of how research is conducted are naive. The second reflects the other side of the self-transformation into a science teacher coin – that of becoming an educational practitioner. This involves adoption of education, and therefore social science, research practice.

To science trainees and new science teachers, educational research can appear messy. They discover clashes of thinking, educational theories and research practices that appear complex and perhaps rather vague compared to the collection of 'hard' data involved in much SCIENCE research. Inductees may see social science as 'wishy-washy' and lacking in the rigour of the scientific method.

However, deciding that the methods of SCIENCE are superior and should be applied to classroom enquiry is unlikely to prove effective in taking on the professional identity of a science teacher.

Consider, for example, the way trainees may go about undertaking small-scale classroom research as part of a coursework requirement for a PGCE. Commonly, their initial research designs reflect the 'fair test' model met in Sc1 of the NC, that is, like a scientific 'investigation'. For example, a trainee may wish to teach two student groups using different treatments to see which is more effective in enhancing learning. One group may have lessons using lots of imagery, and another using physical models instead of visual aids. Students' results on a summative test will, the trainee proposes, identify which method was more effective. Assuming the test is a fair assessment of what is being taught, then – all other things being equal – the better test results indicate the better method. But, of course, all other things are not equal; how could the trainee teach two groups in exactly the same way (apart from the 'manipulated' treatment aspect)? *Teaching* is an interactive process, so is always a function of the *taught*, as well as of the *teacher*. Also, the two classes cannot have been taught at exactly the same times, in the same room, following the same timetable. Even if both classes were meant to be mixed-ability, and were supposedly matched, the trainee cannot be sure that there are no differences in how the students would have learnt about this topic and performed on the test. How would s/he find out, and correct for this?

The 'scientific investigations' model of Sc1 cannot be applied to classroom-based enquiry. Students, and teachers, are complex phenomena, classes are even more complex, and interacting combinations of students and lessons are complex interactions between teachers and classes. Education is a social science, and educational enquiry requires social science approaches.

Part III will set out a rationale for adopting a critical but open-minded, perspective on education theory and research. In Chapter 11 we provide general conceptual tools for becoming users of educational ideas. We follow this, in Chapter 12, with a review of some major aspects of educational research, and in Chapter 13 we encourage trainees and inductees to become creators of small-scale, classroom-based research.

Chapter 11

Being a reflective practitioner

Even while they teach, wo/men learn.
(Seneca ('the Younger') *c*.4 BC–AD 65)

INTRODUCTION: THE ROLE OF CRITICAL REFLECTION

In our Preface we claimed we would explore two statements that inductees to science teaching might find strange, if not challenging:

science teachers are not scientists;
school science is not SCIENCE.

We related these claims to one of the key themes of this book, that of the inductee to science teaching being in the process of adopting a new professional identity. Our argument is that although a new entrant to science teaching is clearly *looking to* take on the identity of science teacher, that *of itself* does not bring about the process. An inductee cannot realise initially what is involved in the professional transformation. If we accept that becoming a science teacher involves fundamental changes in perspective and professional identity, then a logical consequence is that these changes can not be fully appreciated in advance. In the past, this process has been something occurring over an extended period of time without much explicit exploration. In our view, this process needs to be explicit and actively engaged with, if an inductee is to achieve a full and relative speedy transformation.

To fully take on the new professional identity, an inductee needs to actively engage with questions such as:

- 'what is school science, and how is it positioned in relation to what professional scientists do?'; and
- 'what is the essence of being a science teacher, and what are the underpinning philosophical and theoretical frameworks that support science teaching?'

We are attempting to explore questions such as these for readers through this book. We do not claim to have complete and definitive answers; rather, we have provided the reader with our own views and beliefs in the hope that most readers will generally come to agree with much of what we say but, perhaps more importantly, that our readers will at least become engaged with the issues. What we are looking for is to provoke critical reflection on the key issues: *thinking about* what we are suggesting, and *exploring*, rather than just accepting, our positions.

The attitude of being a critically reflective practitioner is essential to acting as professional in such a contested field as science teaching. There are plenty of powerful, and influential, organisations that will want to tell you how to do your job: what to teach, how to teach it and how to assess the students' learning. Sometimes, but not always, the advice will be sound. On occasions, you will be told what to do by authorities such as the DfES, QCA and Ofsted. As a professional teacher you need to be able to judge which advice is worth following, which guidance it is better to ignore and which requirements you should challenge for the benefit of your students.

THE SCIENCE TEACHER AS A CONTINUOUSLY DEVELOPING PROFESSIONAL

We explored the meaning of school science and the nature of the science teacher's expertise in Part I. School science is a socially constructed entity meant to *reflect* SCIENCE, but can never *be* professional SCIENCE. This leaves open the extent and ways in which science as a school subject should and can reflect professional activities in the sciences. A reader who has read this far will appreciate that:

- we believe there *are* ways school science can give an *authentic* taste of subject matter and SCIENCE processes;
- we see part of a science teacher's role (collectively and individually) as finding ways to transform SCIENCE into school science to give something authentic and yet educationally appropriate.

Given the various political influences that impact upon school science, especially through the formal curriculum and the constraints of external qualifications, this process of creating authentic and educationally appropriate school science is a major professional challenge. Two things are clear: first, the individual teacher cannot and should not be acting alone in this process and second, a science teacher is unlikely ever to be able to claim they have got it 'totally right'. We discuss both these briefly.

A teacher cannot act alone

A key part of taking on the professional identity of science teachers is about finding the balance between individual and collective action. A science teacher has a highly responsible job and cannot shed the responsibility for making decisions in planning and executing teaching. That same teacher is a member of a department, a staff within the institution and wider professional groups. Many readers of this book will aspire to take on additional responsibilities in a science department, perhaps eventually becoming a Head of Department. At that time the teacher starts to take on a leadership role, advising others on many of the key issues raised in this book (cf. Parkinson, 2004).

As we pointed out in the introduction to Part I, all educators take on multiple roles in their teaching positions. As a professional teacher, we may seek to contribute to and benefit from our membership of the GTCE and/or one of the trade unions and that is quite appropriate. On other occasions, we may seek advice, or offer to contribute, as a *science* teacher to another professional reference group such as the ASE or one of the learned societies. The learned scientific societies all recognise the role of education in ensuring the public is informed about their work, helping to safeguard future recruitment into science degrees and scientific careers (see Box 11.1).

A science teacher can never get it completely 'right'

A second conclusion from the first two parts of this book is that a science teacher will never be able to claim s/he has totally 'got it right'. The complexity of the task, the ever-changing demands and expectations on teachers mean that there is always room for improvement and a need to accept change. Any professional is expected to engage in continuing professional development (CPD).

BOX 11.1 EDUCATIONAL ACTIVITIES OF SOME OF THE SCIENTIFIC SOCIETIES

The Institute of Biology

> We have made major contributions to the evolution of the QAA as well as the QCA. The British Biology Olympiad and our entries to the International Biology Olympiad are now well established and highly successful. The development of an acceptable scheme for Continuing Professional Development continues to be a high priority ... The Institute of Biology has a strong tradition of supporting science educators.
>
> (From the Institute website, www.iob.org/, accessed 11 April 2004)

The Institute of Physics

> Education in science and physics is a major concern of the Institute. It is, therefore, involved in policy discussions and policy implementation at all levels – from primary education to postgraduate and post-doctoral training. The Institute is involved in debate on the nature and content of the physics component of the Science National Curriculum at secondary education level ... The Institute's Post-16 Initiative is a radical, forward-looking initiative to revitalise physics post-16, attract students, support teachers and influence the direction of future syllabuses.
>
> (From the Institute website, www.policy.iop.org/Policy/education.html, accessed 11 April 2004)

The Royal Society of Chemistry

> The RSC's educational activities provide information and training opportunities for both students and teachers. The RSC is extremely active in determining the future of chemical education, seeking to influence Government by submitting evidence to Parliament and anticipating developments in education policy.
>
> (From the Society website, www.rsc.org, accessed 11 April 2004)

Teachers undertake CPD by attending in-service courses in schools, LEA centres, or by travelling to national meetings like the annual meeting of the ASE, or the BA Science Festival. Courses can be expensive, and will require a substitute teacher if they occur during term time. The expense means that schools usually expect the aims of a course can be shown to match with individual, school or departmental needs identified in personal and/or institutional development plans. This is a healthy attitude as the teacher appreciates what is to be gained from the process. However, the sense of courses as being something done *to*, rather than *by*, the teacher prevails in some cases. The agenda for training is more commonly being set externally, such as through the KS3 strategy or various other government initiatives, leaving limited money and time for training in areas of *personal* priority.

Taking responsibility for professional development

Access to in-service courses has long been seen to be an important resource for teachers, but since the notion of CPD as a professional *requirement* (rather than an entitlement) has become widespread, systems have arisen to provide mechanisms to identify needs, provide inputs, and record outcomes. Although we support this in principle, we also urge caution, as there is a danger of teachers' CPD being hijacked and taken out of their control. This contributes further to the centralisation of the role of teachers, reducing their individual professionalism further.

A teacher should remember that it is *their* professional skills and knowledge that are being developed. If the teacher identifies subject knowledge in particle physics as a priority deficit that needs to be addressed or recognises that a year 9 unit of work on chemical reactions is not successful, then CPD should be addressed to these areas. It is not acceptable to be sent, instead, as the department's delegate on some unrelated course because no one else wants to go and it is your turn to have a 'day out'. Government has priorities for the education system, such as more effective application of ICT in teaching; inclusion of students with special needs; and meeting the needs of gifted students. There is nothing wrong with the government having initiatives and providing support in these areas, but this very visible and often well-resourced activity should not be to the exclusion of meeting CPD needs identified by individuals and departments. In a worst-case scenario, the science teacher may think they have very limited external support for professional development in those areas where they believe it is most needed.

However, almost all schools now have appraisal schemes for all staff, at which professional development needs and career form part of the discussion. This process starts with the writing of a 'Career Entry Profile' at the end of an ITT year. This is taken forward to the inductee's first post and is used to guide professional development during the NQT year. Many schools are now very aware of the needs of inductees and work hard to provide a good environment in which to support this stage of acquiring the professional identity of a teacher. Regular appraisals of staff training needs form part of schools' on-going development plans. Participating in these can be beneficial to developing a sound, realistic and clear professional development plan. We give further advice about this below.

A personal, professional development plan

We suggest that taking responsibility for one's professional development involves building one's own development plan. When departmental or school support is available to help build, or respond to, the plan, so much the better. Ultimately though, each of us needs to take a view about our intended professional goals and the development path needed to achieve them. As careers proceed, professional development needs may be associated with aspirations towards new posts and/or challenges. For the moment, we address inductees and assume that the development plan primarily concerns aspects of a teaching role.

The starting point for a plan is lesson evaluations (see pp. 206–7). These provide a source for identifying what should be done better, why it falls short and what changes are needed. For an inductee, many short-comings will relate to inexperience, so the process of evaluating teaching by critical reflection will clarify how things may be done differently in future. Recognising areas requiring more than just further teaching experience, such as a lack of subject knowledge, development of ICT skills or a need for more pedagogic content knowledge is a good starting point.

After identifying areas for development, avenues for support can be explored. A mentor, head of department, or the senior member of staff with responsibility for staff development may be able to help. If not, there may be support available in the form of courses offered through the ASE or the learned societies. Sometimes these are available at subsidised rates to members.

If subject knowledge is an identified issue for development, then books and internet sites may provide useful sources of information and support. Bearing in mind the problems associated with working outside one's specialist area (see pp. 95–100), a more experienced teacher with specialist knowledge could become an informal 'tutor'. Joining one of the science teachers' discussions lists (see p. 152) can also be a good source of support. For example, consider the case of a biology specialist teaching physics topics and aware that they are not happy about their own background knowledge in some areas. If, after reading-up the topics, there are still doubts about fully understanding key points (and assuming there is no physics specialist to help, which may be why the biology specialist is teaching this material), the discussion list may be a helpful source of advice. An honest posting identifying the sticking points or areas of uncertainty is likely to elicit a number of responses from more experienced colleagues offering advice.

USING RESEARCH EVIDENCE AND RESEARCH FOR PROFESSIONAL DEVELOPMENT

Experienced teachers can offer advice on pedagogic issues as well as on subject knowledge. Pedagogic problems are often specific to the context in which the inductee works. Moving schools will often bring the same subject knowledge problems, but different pedagogical issues. Development of pedagogic skills can require a more proactive approach.

A place for education research

A reflective practitioner can resource his/her professional development about pedagogical issues by accessing and learning from the vast literature about teaching and learning science. Box 11.2 describes common *approaches* used in educational research. Chapter 12 explores issues relating to use of research-base evidence to inform practice, and describes some of the actual *techniques* used in educational research (see Box 12.1, p. 226). The next step is to actually undertake small-scale classroom research (we discuss this in Chapter 13).

BOX 11.2 SOME APPROACHES TO EDUCATIONAL RESEARCH

Hypothesis testing – there is a good deal of educational research following a 'positivist' paradigm which, like the natural sciences, seeks to test hypotheses. However, the complexities are much greater than are usually the case in science, controlling conditions is impossible and the research 'subjects' can be influenced by research in unexpected ways. This type of research depends on careful selection of samples, to ensure control and treatment groups are sufficiently similar and designing the research to look for *statistically significant* differences between different groups which are *unlikely* to be the outcome of chance.

Surveys – educational research explores the opinions, beliefs and attitudes of students, their parents or teachers. If the research is used to inform an organisation such as a school, surveying all those involved may result in a high return rate. Surveys reporting the views of specific groups such as 'teachers', 'science teachers' or 'sixth-form students of biology' are much more difficult, as responses from a small proportion of the target group may be obtained. If the results of such surveys are to be meaningful then reliable sample size and sampling techniques are needed. For example, a survey of 'science teachers' should include a representative range of teachers in terms of the factors thought to influence the attitudes being explored: this could include age, gender, specialism, years of teaching experience, type of school and level of qualifications.

Case studies – the technical difficulties of designing and carrying out research to test hypotheses are avoided in educational research that focuses on a detailed understanding of an individual case. This type of work often involves obtaining data from a range of techniques to build up an in-depth description of the 'case'. Approaches may be influenced by ethnography, that is, exploring the meaning that classroom acts have for teachers and/or students, rather than attempting to judge them objectively. Case studies can provide detailed and authentic accounts but they are not generalisable, the reader must judge their relevance to their own students and/or teaching.

Science teachers are likely to find the idea of being teacher-researchers an attractive proposition. We close this chapter with a note of caution, though. Being a scientist does not *automatically* equip one to understand, interpret, evaluate and undertake educational research. Education is a social science, a fact reflected in the methodologies, data and outcomes. To understand, and undertake, educational research means working within *social* science paradigms, which may seem non-rigorous (and even subjective), compared with approaches in the natural sciences. Nevertheless, there is no doubt that, despite the complexity of the subject matter, making sense of the data that educational research often generates (e.g. see Box 11.3) can be immensely satisfying and rewarding.

Accessing education research

Education research is readily accessible on one hand, but needs to be sought out on the other. Science teachers who are ASE members automatically get a reliable, readable, up-to-date source of research in the form of *School Science Review*, the association's main journal published five times a year. Other strategies for accessing research are described in Box 11.4.

Science education research is published in a range of journals. Those with the highest 'status' include the *International Journal of Science Education (IJSE)*, *Science Education (SE)*, *Journal of Research in Science Teaching (JRST)* and *Research in Science Education (RSE)*. There are journals which feature science education with a specific focus such as *Chemistry Education Research and Practice* (on-line at www.rsc.org/uchemed/cerp.htm), the *Journal of Science Education and Technology*, *Research in Science and Technology Education*, the *International Journal of Science and Mathematics Education*, and the *Journal of Science Teacher Education*. Other periodicals with research input, but which are not exclusively research-based, are *Journal of Biological Education* (published by the Institute of Biology), *Education in Chemistry* (published by the Royal Society of Chemistry) and *Physics Education* (from the Institute of Physics Publishing). In addition, *Studies in Science Education* publishes longer, review-type articles once or twice a year.

Meetings of research-based organisations are held mainly internationally. For example, the European Science Education Research Association (ESERA) and the International Conference on Chemical Education (ICCE) meet biennially; the North American Research in Science Teaching

BOX 11.3 DRAWING CONCLUSIONS FROM EDUCATIONAL RESEARCH

Meta-analysis – in natural science, experiments are often reproducible, leading to experimental results being replicated and corroborated. In education, there are often many similar studies with less clear-cut findings. This may be because the studies have been undertaken in countries with different cultural and education contexts, perhaps with students of different ages or ability ranges. Even when these factors are similar, different teaching styles or levels of student motivation may lead to variation in outcomes. Meta-analysis attempts to synthesise the findings of many related studies to see if there are overall patterns that can inform educational practice. An example of this type of study is Black and Wiliam's (1998) review of research into the effectiveness of classroom assessment.

Grounded theory – one way of bridging the specificity of case studies and the desire for more generally applicable conclusions, is an approach known as grounded theory (Charmaz, 1995) developed by the sociologists Glaser and Strauss (Glaser and Strauss, 1967; Glaser, 1978; Strauss and Corbin, 1998). Grounded theory works from individual cases, but goes beyond description. By a method of 'constant comparison', and the 'theoretical sampling' of further cases, grounded theory approaches allow researchers to develop theories suitable for later hypothesis testing. This has much potential in science education (Taber, 2000b).

We invite you to put aside any prejudice you may have about social sciences. Educational research is not like natural science research, but there are good reasons for this (Cohen *et al.*, 2000). Understanding and accepting the way educational research works and engaging with this new approach to enquiry, may be quite a difficult transition. It is, however, the final major step in taking on the new professional identity of a science educator rather than being a scientist.

(NARST) and Australian Science Education Research Association (ASERA) meetings take place annually. Key figures present their latest work at such meetings or reflect on trends and developments in the field as a whole. Teachers are welcome as members and to play a full role in these organisations. The ASE exists to help teachers in all aspects of their practice, including research.

BOX 11.4 WEB-BASED STRATEGIES FOR ACCESSING SCIENCE EDUCATION RESEARCH

By topic: go to a database such as ERIC. These are often accessible through libraries. Enter the topic in the search engine.

By journal: go to the website for the journal.

By author: enter the author name in a general search engine. This may produce the author's own webpage from which papers may be available. Alternatively, go to a database such as those listed above and search by author name.

By specific reference: this can be obtained either from a journal website, database or search engine.

Papers can be purchased from journals at a relatively small fee. Sometimes the fee is waived for students.

A non-ICT-based approach is through printed literature. Many good bookshops have a selection of science education texts and we have recommended many throughout this book. An excellent introduction to aspects of science education research is Judith Bennett's *Teaching and Learning Science: a guide to recent research and its applications* (2003). We use some of her material in the next chapter. Other texts, as we have cited earlier, focus on aspects of science education research such as students' alternative conceptions, probing understanding, scientific literacy, explaining science and making meaning.

Government websites for the QCA, DfES and Ofsted, as well as those for the RSC, IoB and IoP (and other related societies, such as for the engineering institutes) also feature research, although this will usually be in the specific areas targeted by these organisations. University education departments have libraries usually accessible to visitors, as well as websites listing publications, offering research papers for downloading and other snippets. Of course, pursuing research interests through a higher degree such as an MTeach, other Masters or Doctorate study will inevitably bring about involvement in education research at a high level. Funding to finance small-scale research projects can be obtained from the learned societies, while the DfES has run a research scholarships project for several years.

Another new source of information, professional development and access to research will be the Science Learning Centres. A project funded by the DfES and the Wellcome Trust, this will provide nine regional and one national centre offering professional development for science teachers of all levels of experience. Some centres are open at the time of writing and others will be opening during 2005. Information about the centres and the courses currently on offer can be obtained from the main website at www.sciencelearningcentres.org.uk/ (accessed 1 August 2004).

Developing an interest in education research can open many doors and enhance a teaching career. There is potential to make international contacts through the community of science education researchers worldwide – a good starting point for this is the European Science Education Research Association (ESERA). Both the authors began their participation in education research quite early on in their careers. One of the authors writes:

> At the end of my PGCE I was aware of having barely dipped a toe in the ocean of education – rather like taking a driving test, passing it and then finding out what life on the road is really like. I later took a Masters degree which introduced me to science education research methods and I undertook a small-scale study of children's learning of the kinetic particle theory. The range of ideas, mainly incorrect, that 11–12-year-old children expressed when invited to explain their thinking about a topic I thought I had taught well was astonishing. I started teaching some lessons using diagnostic questioning strategies to explore explicitly children's thinking about ideas such as particles and chemical reactions. I found that children's motivation improved as they appeared to learn more meaningfully through being made to think! Inspired, I continued to a doctorate study of 16–18-year-old students' learning of basic chemical ideas, then returned to teaching. At this point I changed my teaching significantly – I was far more aware of students' learning patterns and how I could impact on their learning by what I said and did. I became much more skilled than previously at getting 'Aha' moments [see Chapter 9] and I developed a lot of my own resources for developing meaningful learning. I achieved excellent exam results during this period. My involvement with science education research I think made me a much more effective and reflective teacher.

The other author recalls:

> I remember reading paperbacks with titles such as 'The Psychology of Learning Mathematics' and 'The Psychology of Study' when I was

studying for my A levels. I bought the books to help me as a student, but found them fascinating. I also really enjoyed reading about topics such as learning theory and the social psychology of the classroom during my PGCE. However, I was aware during my training year that too much of my energy and time needed to be focused on lesson preparation and developing basic classroom skills for me to be able to properly explore how these ideas could be used in my own teaching. I told myself that once I had more experience, competence and confidence, then I would revisit some of the educational theory. During my third year of teaching I enrolled on a distance-learning course reading for a Diploma in the Practice of Science Education. This was assessed by a series of 21(!) written assignments, many based on evidence collected through small-scale research or development activities in my own classes. For example, I remember being amazed to find that some of my lower secondary (KS3) students believed that condensation was made though heat mixing with 'cold'. The diploma course ignited an interest in educational research, and in my next school I enrolled for a Master's level course, researching girls' underrepresentation in physics classes. Next I moved to work in FE and started a doctoral study exploring an aspect of A level students' science learning. Throughout this period of time my classroom research investigated, and informed, aspects of my professional practice.

We can both attest to the sustained improvement in our pedagogic practice through this work, as well as to the varied opportunities that become available to those who have taken this extra step.

THINKING ABOUT PRACTICE

- Consider your strengths and weaknesses as a classroom practitioner. What would you prioritise as your three most important needs for professional development?
- If you are considering teaching as a long-term career, think about where you would like to be in five and ten years' time. What professional development may be needed to help you achieve these aims?
- What are the advantages and disadvantages of being involved in education research?

Chapter 12

Using research-based evidence in teaching

Education is a sieve as well as a lift.
(Sid Chaplin)

Science education research has much to contribute to classroom practice. Developing the habit of utilising relevant findings will help a teacher reflect on and enhance his/her skills. This chapter provides an introduction to science education research and a survey of four areas indicating how each has been, or may be, used in teaching.

INTRODUCING SCIENCE EDUCATION RESEARCH

One of the authors had a first job in a research laboratory in the early 1980s when molecular biology was identifying and characterising genes responsible for specific diseases. Experiments involved multi-stage techniques called 'Southern blotting', 'plasmid preparation', 'gene library screening' and so on. Most required a 'control' against which other results could be compared. Reading scientific papers and attending talks by leading molecular biologists of the day helped create a sense of immersion in a scientific world based on generating hypotheses, setting up controlled experiments and analysing data against 'standards' or by making comparisons with other, earlier data. To continue the cycle, 'we' (that is, the research group) worked out what to do next in the light of experimental findings.

This pattern may be familiar to anyone who has worked in SCIENCE, at least in a research capacity. Based on these experiences, scientists may have definite ideas about what constitutes 'research' which may influence opinions about science *education* research. Let us now consider science education research with this in mind.

The science education research tradition developed since the 1970s and practised internationally draws almost entirely on *social science* practices. The process begins with a research question framed as a problem or hypothesis, for example:

What ideas and understanding do 16–17-year-old chemistry students have about chemical equilibrium?

(Banks, 1997)

Are students' attitudes to science indicative of their general attitude to school and is there a gender bias in such attitudes?

(Smith, 1994)

How can teachers use formative assessment to support pupils in setting effective short-term targets for improvement in their work?

(Wall, 2002)

These examples are questions generated by classroom teachers. To answer them, a researcher chooses an appropriate strategy or technique for getting data. Box 12.1 gives the main choices.

Most of these techniques differ markedly from what a scientist may consider 'research'. In common with many scientists entering education, you may regard only 'experiments' as 'research'. In contrast to 'hard' SCIENCE, it is very difficult to set up controlled experiments in education, as there is no such thing as two (or more) identical classes. To obtain data in science *education* research requires that other methods are applied.

The research strategies in Box 12.1 often generate data reported in one of two modes – qualitative, involving mainly words, and quantitative, involving mainly numbers. Qualitative data provide a rich overview of respondents' thinking, such as ideas about a science topic, or views about a new initiative. Qualitative data analysis looks for patterns in responses, perhaps exploring the extent to which views are common among different groups of respondents, or the frequency with which a certain answer occurs within a group of children. When working in a classroom setting, qualitative data can give startling insights into the impact of teaching on learning, unexpected views about assessment or a strong impression of how a policy is being implemented.

BOX 12.1 STRATEGIES AND TECHNIQUES USED IN EDUCATION RESEARCH

Questionnaires A questionnaire is an effective technique used to collect data in large- (that is, hundreds) or small-scale (10–100) studies. Diagnostic questions exploring students' thinking about a specific idea or concept in science may be used. These normally comprise a 'stem' of information, followed by a factual question followed by an invitation to explain reasons for the answer.

Interviews One-to-one recorded interviews are popular for collecting high-quality but small-volume data. These can expand on or verify information gathered using another technique. Interviews are also useful as a primary data gathering method to probe directly students' or teachers' views or ideas about a topic under investigation. Data quality can be very high but the number of interviews normally practicable within time constraints is often limited.

Case studies are used widely to explore one issue from a wide range of perspectives. A case study may focus on one school, using a range of techniques to collect data from staff, governors, students and parents. The data can be comprehensive, providing a thorough picture of the issue. Generalising findings to other situations can be problematic, so researchers need to present findings carefully.

Focus groups These have become a fashionable way of getting information about an issue from a group of individuals simultaneously. A focus group will discuss specific questions or a topic, perhaps using prompts. The researcher directs the group but does not contribute. A focus group may collect a large amount of data efficiently, so may be preferred in some circumstances to interviews.

Observation In this technique the researcher is present in the classroom, recording what is happening using notes, a specifically designed 'tick' sheet, video camera or tape recorder.

Action research is used frequently by classroom teachers interested in developing their own practice. The researcher is part of the study – for example, as a teacher teaching a sequence of novel lessons then exploring the impact of these on students' learning. Evaluation of the impact may lead to further changes in practice, so action research often has a cyclical feature.

Ethnography has its roots in social anthropology. The researcher becomes a member of the group under investigation, for example, joining the staff of a school to study the implementation of a new curriculum or behaviour strategy, then explores with group members how and why they acted as they did. Such studies can provide a fascinating insight into a specific situation, but generalising findings to wider audiences can be problematic.

Experiments involve setting up a hypothesis or problem in a similar way to 'normal' science using controlled and non-controlled variables. The effect of change on one or more variables is the experiment. This technique is used infrequently because finding controllable variables in education is problematic. Also, ethically it is hard to deny a group of children a positive learning experience because they were the 'control' group.

Surveys Major national and international groups use surveys to explore issues on a large scale, possibly involving thousands of students said to be a 'representative group'. A survey often uses questionnaires (see above). The scale of a survey often makes mathematical analysis of data possible and statistically reliable. Comparative studies often involve surveys, as one questionnaire is issued to different groups.

Document review A historical study or a study of policy is likely to require a review or survey of relevant documents. A researcher will sift through for information to help answer a research question, then may supplement the data with interviews or questionnaires.

Quantitative data are generated most frequently by large-scale surveys or questionnaires involving several hundred participants. Responses are subjected to statistical analysis and findings are expressed numerically rather than descriptively. The purpose of statistical analysis is often to assess how likely it is that the outcomes were obtained by chance, or if any other factors may have been an influence. Other aims are to provide evidence of students' learning in numerical terms, perhaps looking at year on year changes, or to make comparisons between groups of students.

Qualitative data have the disadvantage of appearing vague and perhaps inapplicable outside the situation in which they were obtained. Quantitative data, when presented alone, give bald figures, ignoring the richness of opinion and insight. Quantitative data can seem more reliable and solid at face value. In reality, studies often combine both types of data.

The issues of reliability and validity are important in education research generally. Reliability is achieved if the same result is obtained time after time using the same technique. Validity is achieved when data can be judged as measuring what they are claimed to be measuring.

A science education research study will often state a range of findings from the data. The answers to the research questions are normally given at this point. The findings are often followed by implications for practice. These last are perhaps the most significant – for example, how, having done the research, might the teaching of a science topic be improved? What improvements could be made to setting or grouping of children given gender differences highlighted by a research study? Is a new laboratory behaviour policy proving effective, or are changes in implementation necessary?

Having outlined science education research, let us now continue to survey the field. This is necessarily brief but, it is hoped, sufficient to encourage a reader to seek more information.

RESEARCH ABOUT CHILDREN'S LEARNING OF SCIENCE

Many areas of science education have been the focus for research studies. This review will introduce research in four areas – children's ideas and their learning of science; cognitive development and children's learning of science; strategies for teaching science; and practical work in science.

Research based on a constructivist theory of learning

Research about children's understanding of science was prompted by curriculum developers in the 1960s who advocated a 'discovery learning' approach to science. This involved children 'acting as scientists' in carrying out experiments. By the 1970s limitations of this approach were apparent. Questions arose about the extent to which children could 'discover' key principles in science for themselves in short time spans, given their prior knowledge, experiences and limited practical skills (e.g. Driver, 1983). The extent to which new learning is influenced by prior experiences gave rise to a new way of thinking about learning in science education, called 'constructivism'. Work produced under the constructivist umbrella (or paradigm) has been the single largest area of science education research.

The psychologists George Kelly and David Ausubel contributed significantly to the development of constructivism. Kelly (1955) described a 'personal construct' theory in which people make sense of their environments by making theories for themselves to explain their experiences. New experiences are then tested against these theories. The theories are reviewed or changed in an on-going construction/reconstruction process. Ausubel, meanwhile, argued for 'meaningful learning' (see Box 6.2, p. 124) to take place in classrooms (Ausubel, 1968). He said that the key was to 'ascertain what the learner already knows and teach him accordingly'. Both these theories contributed to dismissal of the 'blank slate' notion that learning involved receiving information transmitted directly. Instead, the knowledge and experiences children acquire prior to (or perhaps as a result of) teaching were recognised to impact significantly on how they comprehend new material.

Rosalind Driver and Beverley Bell (1986) provide a short summary of the constructivist view of learning:

Learning outcomes depend not only on the learning environment, but also on the knowledge of the learner.

Learning involves the construction of meaning. Meanings constructed by students from what they see or hear may not be those intended. Construction of a meaning is influenced to a large extent by existing knowledge.

The construction of meaning is a continuous and active process.

Meanings, once constructed, are evaluated and can be accepted or rejected.

Learners have the final responsibility for their learning.

There are patterns in the types of meanings students construct due to shared experiences with the physical world and through natural language.

(pp. 353–4)

Much research in science education has been undertaken under the premise that children construct their own meanings and understandings from their everyday experiences. Large-scale studies such as the Children's Learning in Science Project (CLISP) carried out in the UK and the Learning in Science Project (LISP) in New Zealand (Osborne and Freyberg, 1985) have contributed to a large base of documentation on (mainly 9–16-year-old) children's understandings of science topics such as electricity, chemical reactions, photosynthesis, forces, heat, magnetism and evolution.

Using the research in practice

According to constructivism, children's ideas are often formed, at least initially, through informal learning, so their thinking may differ significantly from accepted scientific viewpoints. The research suggests that children's understandings exhibit a range of characteristics, such as:

- being difficult to change and persisting even after teaching;
- confusing several scientific ideas together in the same misunderstanding;
- developing from observable phenomena;
- being specific to one context, so it is all right to think differently when another similar situation arises.

As a science teacher, being aware that children are unlikely to learn a topic as intended is an important step. How best to teach science knowing that children may have incorrect ideas is the next. Researchers have considered the implications for teaching. An accepted strategy is to begin teaching a topic by finding out what the children already know – this is referred to as 'elicitation'. Ideas for how to prepare for this can be found in White and Gunstone (1992). Next, children need an activity introducing the expected idea in a way that challenges their original thinking. Children enjoy stimulating discussion, thinking about a demonstration, or considering how their ideas and a 'scientist's' ideas match or clash. Suggestions may be found in Wellington and Osborne (2001) and Ross *et al.* (2000). Barker *et al.* (2002) provide materials for eliciting and teaching ten 'difficult' topics in 14–16 science, and the Royal Society of Chemistry have provided probes for exploring students' ideas in core chemical topics over the 11–19 age range (Taber, 2002b – materials available at www. chemsoc.org/networks/learnnet/miscon2.htm, accessed 1 August 2004).

Posner *et al.* (1982) suggest that for a new idea to be accepted, it must appear to be *intelligible*, that is, understandable by the audience; *plausible*, that is, it must make sense by offering explanations; and *fruitful*, that is, it must offer more than the original. This process is unlikely to be completed in a 45-minute period – children will need repeated opportunities and time to make changes to their thinking.

Criticisms and developments of constructivism

As a research paradigm evolves, inevitably researchers also reflect on the paradigm itself. Hence, constructivism has been criticised and developed. One major criticism, levelled by Robin Millar (1989a) is that constructivism has become linked invalidly to a particular model of teaching. As a child constructs ideas in his/her own head, this process is independent of any instruction model. Therefore, constructivism cannot imply that any one method of instruction is preferable. Millar continues to suggest that active involvement and engagement with science is more likely to create a positive learning environment within which scientific ideas may be learned. A second criticism is that constructivism focuses on individuals at the expense of social learning contexts. This is because the original forms of constructivism arose from 'personal construct' theories in which the individual has the central place. In a 'real' environment, the context and setting for learning is, of course, significant.

Constructivism has, therefore, evolved towards a 'social constructivism' in which the wider social and individual processes are both acknowledged as important in the construction of knowledge. Researchers have focused more on interactions in the classroom rather than on individuals, while work has also been done on developing science curricula that seek to create positive and stimulating learning opportunities.

Research about cognitive development and children's learning in science

The cognitive development theory developed by Jean Piaget and his co-worker Bärbel Inhelder in Switzerland between the 1920s and 1960s is, perhaps, the best-known and most studied theory in relation to science education. Their theory (Inhelder and Piaget, 1958) claims that all children pass through common stages of development, and that in each stage they exhibit similar behaviour and thinking skills. For science (and mathematics) education, the final two stages are the most significant. These are called *concrete operational* and *formal operational* thinking. A child who can think in a concrete operational way will, for example, be able to 'conserve' the amount of liquid when poured from a short, wide glass into a tall thin one, in other words, realise that the amount stays the same. Concrete operational thinkers can also classify objects systematically and look for sub-categories. Children's thinking in this stage is based on their analysis of real, that is, concrete, events. At the

231

formal operational stage, children are able to think in terms of abstract ideas. They can formulate hypotheses and test these using variables. They also understand mathematical principles such as ratio and proportion.

Piaget suggested that movement through the stages depends on children maturing with age, the environment in which they live, the activities they do and the social interaction they experience. Science education researchers have investigated the role of activities in children's cognitive development as a route towards accelerating progress to formal operational thinking.

Although Piaget's theory has won wide acceptance as it reflects people's general experiences of child development, there have been significant criticisms. Margaret Donaldson (1978) tried Piagetian-type tasks with very young children, but situating them in a context appropriate to the children's ages. She found that the children were able to think in a much more sophisticated way using these adapted tasks than Piaget's theory predicted, discrediting the age maturation factor for progressing through the stages. Novak (1978) pointed out that individuals can use different Piagetian thinking in different circumstances – for example, someone may be a concrete operational thinker in one context and use formal operational thinking in another. Other research also suggests that children's performance levels on Piagetian tasks are influenced by their prior experiences. Nevertheless, Piaget's theory has had a significant impact on the organisation and approaches used in teaching science, particularly during the 1970s. The theory has provided the focus for a major research study and teaching programme seeking to enhance children's cognitive development.

Accelerating children's cognitive development – the CASE project

Philip Adey and Michael Shayer have been the main UK-based researchers associated with cognitive development since the 1970s. Their work has resulted in the development of Science Reasoning Tasks (SRTs, Shayer and Adey 1981) to assess children's levels of cognitive development; analysis of science curricula against Piaget's stages of development through using Curriculum Analysis Taxonomy (CAT) (also Shayer and Adey 1981); and activities to accelerate progress through stages (the Cognitive Acceleration through Science Education or CASE project, Adey et al., 1989).

Their work suggested that the content of science curricula used in the 1970s and 1980s in the UK was too abstract in nature for most 14–16-

year-olds to understand. This is because their SRT and CAT results indicated that most 15–16-year-olds have not reached the formal operational thinking stage needed to understand the most abstract science topics. As a result of their findings, Shayer and Adey began to investigate activities that may be used to accelerate children's cognitive development. With other workers, they produced *Thinking Science* (Adey *et al.*, 1989), a programme for 11–12-year-olds comprising a set of activities each requiring 70 minutes to deliver. Each *Thinking Science* or CASE activity includes:

- an introduction, generating *conceptual readiness*;
- the presentation of problems which create *cognitive conflict*;
- stimulation of *metacognition*, that is, children reflecting on their own thinking;
- *bridging* of thinking strategies to make connections with other areas.

In developing the programme, Shayer and Adey used a controlled experiment strategy involving 24 representative samples of 11- and 12-year-old children in nine different schools. Classes were randomly assigned as 'control' and 'experimental' groups. Initially, *Thinking Science* lessons were given once every two weeks for two years, with teachers being trained at workshops. At the end of the study, ten control and experimental classes remained, comprising 208 and 190 children respectively. The findings assessed the effects of *Thinking Science* lessons using Piagetian Reasoning Tasks (PRTs – a version of the SRTs from their earlier work) and a science achievement test. Immediately at the end of the intervention, the results showed that the experimental groups performed at a higher level on the PRTs than the control groups, but there was no difference in the science achievement test results, even though as a result of the *Thinking Science* lessons 'experimental' children had received 20 per cent fewer 'normal' science lessons. However, the GCSE results for the experimental groups were significantly higher in mathematics, science and English. These examinations were taken two or three years after the intervention (Adey and Shayer, 1993, 1994).

Criticisms of CASE

These striking results received wide publicity at the time of their release in 1991. Subsequently, many schools adopted the *Thinking Science* programme (often called CASE), to improve their science examination results. 'CASE-training' is available through a network of coordinators. The Cognitive Acceleration through Mathematics Education ('CAME') scheme and CASE materials for 7–11-year-olds have also been developed.

Despite the acclaim, researchers have regarded the original experiment with scepticism. Jones and Gott (1998), for example, raise issues about the differences in schools' results, suggesting that further work is needed before CASE can be said undeniably to 'work'. Leo and Galloway (1995) claim that *Thinking Science* activities best suit children who enjoy problem solving, while for others the programme is of little or no benefit. As CASE is used in a wider range of schools, any benefits in children's learning will gradually become apparent – teacher confidence in the materials, good-quality training, and maintenance of motivation are likely to be factors influencing CASE's long-term success.

Using the research in practice

Schools that have adopted CASE are already using the outcomes of cognitive development research. The activities link to a specific theory and are designed as an intervention to enhance children's development and subsequent examination performance. However, CASE is not a 'quick fix' – several years elapse between the completion of the *Thinking Science* programme and the taking of GCSEs. The training may be useful for a teacher interested in pursuing cognitive development and how this may be encouraged in the classroom.

RESEARCH ABOUT PRACTICAL WORK IN SCIENCE

Practical work in science lessons is a very common practice, to the extent that UK children often enter a science lesson asking, 'Are we doing an experiment today?'. The British practical work tradition began in the nineteenth century with teacher-led demonstrations, then with children doing their own experiments. This 'discovery learning' or 'heuristic' approach continued for some years, being replaced between about 1900 and 1950 by 'recipe-following' experiments designed to illustrate a specific theory or idea. Discovery learning resurfaced in the 1960s with the Nuffield curriculum development programmes, which fell out of favour with the rise of constructivism (see p. 88).

Practical work in the 1970s–80s became focused towards the process of doing science (see pp. 42–3), leading to projects such as *Warwick Process Science* (Screen, 1986) and *Science in Process* (Wray, 1987). Following Assessment of Performance Unit (APU) reports (APU 1988a, 1988b, 1989a and 1989b) 'investigations', in which children raise a testable hypothesis, carry out a controlled experiment and collect and analyse appropriate data, were adopted as the way forward under the science NC. This approach was considered to train children in the process of scientific enquiry (but see the comments about 'Sc1' on pp. 27–8).

Research about practical work since the 1960s has contributed to the changing approaches over this period. However, aims for practical work have been developed throughout this period and are now widely accepted by many science teachers (see pp. 147–50), almost without thinking. Here we will explore briefly what these aims are and how research views them.

The purposes for practical work – is it as useful as we think?

Practical work is generally regarded as useful in teaching science, perhaps because in some way it matches what scientists really 'do'. Research has found that, in fact, much practical activity may be of limited use as a way of encouraging science learning. We can explore this beginning with a look at five aims for practical work suggested by Derek Hodson (1990):

- to teach laboratory skills;
- to enhance the learning of scientific knowledge;
- to give insight into scientific method and develop expertise in using it;
- to develop certain 'scientific attitudes' such as open-mindedness, objectivity and willingness to suspend judgement;
- to motivate pupils by stimulating interest and enjoyment.

(Quoted in Bennett, 2003: 79)

As a set of general aims, Hodson's list probably meets wide agreement. Lists such as these set very high expectations for the outcomes of practical work and may reflect teachers' views more than their actual (or possible) practice. Let us look at researchers' criticisms of these aims.

In terms of developing laboratory skills, researchers suggest that, in fact, most of these can be regarded as either cognitive processes, such as observing and classifying (Gott and Duggan, 1995) or 'craft' skills such as reading a balance or using a burette (Hodson, 1990). Both types can reasonably regarded as 'laboratory' skills. However, the APU found that many children have difficulties with standard laboratory procedures and use of equipment even after several years of science lessons. Other researchers suggest that cognitive skills 'learned' in science are unlikely to be transferred to other subjects, so science teachers cannot reliably claim that practical work generates widely useable skills.

Little evidence exists to support the aim that practical work enhances the learning of scientific knowledge. For example, Watson *et al.* (1995) found little evidence that practical work enhanced children's understanding of combustion, while Ogborn *et al.* (1996), like Driver (1983), question the notion that observing phenomena develops understanding, as children are unlikely to infer an abstract scientific idea from a set of data. Practical work may serve as a bridge that teachers may use to help connect observations with scientific explanations.

The claim that children gain insights into scientific method through practical work has received extensive criticism, partly due to lack of agreement about what constitutes 'scientific method' (see p. 27). An *inductive* or discovery learning method requires children to focus on what they themselves can find out. In practice, children often carry out practical work with the aim of finding the 'right' answer, or what the teacher had expected to happen. Discovery learning has now been largely replaced as a theory for scientific method by the *hypothetico-deductive* method proposed by Karl Popper (1959). Under this philosophy, scientific method is perceived as involving rigorous hypothesis testing. Millar (1989b) notes that school environments are not the best places for testing hypotheses, so it is not realistic to claim that school science reflects this method.

Criticism has also been levelled against the claim that practical work aids development of scientific attitudes. First, researchers do not agree on what 'scientific attitudes' are (see Box 2.2, p. 37). Next, as a result, there is wide variation in the nature and types of studies on scientific attitude, so general conclusions are difficult to justify. Gardner and Gauld's (1990) review of work in this area suggests that teacher behaviour and the structure of practical work may positively influence children's scientific attitudes, although 'curiosity, openness and a willingness to solve everyday problems scientifically remain hopes' (p. 151).

Motivation remains the fifth aim for practical work. Children generally seem to enjoy practical science, most likely because of the freedom of movement and rest from listening to teachers this brings. Research indicates, though, that children prefer practical activities with clear purposes, that are challenging and provide them with control over what they must do (Watson and Fairbrother, 1993).

Using the research in practice

Taken together, these criticisms suggest that teachers need to take care when planning and using practical work in science. It is easy to set lofty aims for practical work in general or even one experiment in particular, but much less easy to ensure these are realised in practice. A practical activity with clear, unambiguous aims, a challenge and some measure of openness for the children is likely to be valuable, particularly when followed up by a meaningful discussion of the outcomes. Variety, too, is also likely to be beneficial. Using different types of activity will yield different learning outcomes for a wider range of children. However, research evidence relating to practical work suggests that using one of a variety of other active learning strategies may be just as valuable on a wet Friday afternoon as getting out the Bunsen burners.

RESEARCH ABOUT CONTEXT-BASED TEACHING APPROACHES

Context-based teaching means using everyday settings or topics as 'ways in' to science. The approach is thought to help generate and maintain children's interest and motivation by expecting them to see science as being relevant to their lives. The first context-based materials such as 'LAMP' – *Science for Less Academically Motivated Pupils* (ASE, 1978) were developed in the late 1970s for children in UK schools for whom the traditional academic courses were thought to be too difficult. 'Slot-in' materials such as SATIS – *Science and Technology in Society* (ASE, 1986) for use in lessons by all students, followed. Curriculum development then turned to devising full GCSE and A level courses using context-based approaches. *Science: The Salters Approach* (UYSEG, 1990–92); *Salters Advanced Chemistry* (Burton *et al.*, 1994); *Salters Horners Advanced Physics* (UYSEG, 2000c), and *Salters Nuffield Advanced Biology* (UYSEG/Nuffield, 2002) are all available under the UK's public examination system. Today, a number of countries around the world have made context-based materials, mainly for 14–19-year-olds, for example, PLON – the Dutch Physics Curriculum Development project (Eijkelhof and Kortland, 1988) and *Chemistry in Context* (American Chemical Society, 1994).

237

Context-based materials emphasise how SCIENCE and, commonly, its technological applications are used in, or appear to, society. Children's everyday experiences may be used as starting points for discussion, leading to the science idea behind them and aiming to develop children's thinking such that they are then able to understand the context, situation or experience. Context-based courses often include a wide range of active learning strategies, such as research, presentations, discussions, problem-solving tasks and role-play exercises. The phrase 'pupil-centred learning' describes this. Kyriacou (1998) suggests that a number of benefits arise from context-based approaches, including enhanced intellectual stimulation and pupil motivation; development of learning skills such as organisation and communication; and enhanced enjoyment and positive attitudes.

The effectiveness of context-based teaching – evidence from research

Context-based teaching has developed mainly through science educators seeking alternative ways of presenting science in schools. A social constructivist influence can be detected in some major curriculum development projects, but we cannot say that these courses have developed directly as a result of research findings. Hence, context-based teaching has developed differently from the constructivist and cognitive development-based approaches to science teaching discussed earlier. In these cases, research outcomes drove the development of teaching materials and/or strategies. The connection between practical work and research also differs from that between context-based teaching and research. Practical work has long been established as an accepted tradition in science teaching and learning and, as such, has acquired aims that meet tacit agreement among practitioners. Context-based teaching, in contrast, is relatively new and not universally accepted. Perhaps unsurprisingly then, there is, to date, relatively little research about the use and effects of context-based approaches, although their impact on aspects of science education is gradually drawing attention. Here is a brief review of the main research findings so far.

Barker and Millar (2000) and Barker (1994) report a longitudinal study of over 300 Salters Advanced Chemistry students in terms of their understanding of basic chemical ideas. The study found that gradual, 'drip-feed' introduction and the opportunity to revisit and develop

238

chemical ideas proved advantageous in certain cases, for example, thermodynamics and chemical bonding. By comparing 'traditional' and Salters students, the study also showed that the context-based approach was as effective as a standard course at teaching basic ideas.

An evaluation of the context-based Dutch physics course PLON by Eijkelhof and Kortland (1988) showed that students' use of scientific knowledge in handling context-based issues varied depending on the degree of publicity associated with the topic. For example, dumping nuclear waste at sea had been hotly debated in the Netherlands, while food irradiation had not. Although both involve discussion of radioactivity, students used their science knowledge much more extensively in discussing food irradiation than dumping of nuclear waste. In both cases, though, students made statements involving misconceptions.

Judith Bennett has studied the impact of context-based teaching on students' enjoyment of science (Ramsden, 1992; 1997). She found that although enjoyment increased, children sometimes did not regard the activities as 'proper' science, as they were enjoyable. Her later study indicated also that uptake of science beyond the age of 16 was not improved by use of context-based courses. Students' negative images of science were often so deep-rooted that using a context-based approach had no impact.

Context-based approaches in practice

Context-based approaches have mainly arisen from curriculum development projects through the 1990s, so as such are still relatively new. Teachers using context-based materials frequently report increased student motivation and, for 16–19 courses, enhanced uptake of science once the course is finished. The studies completed to date find no negative impacts on student conceptual learning and some enhanced positive effects on the social aspects of learning science. That relatively little research has been done in this area so far suggests that rich pickings may be available for teachers willing to try these methods and evaluate them, perhaps as action researchers working in their own contexts (see Chapter 13).

SUMMARY: USING RESEARCH-BASED EVIDENCE IN PRACTICE

This chapter has introduced science education research and reviewed four areas that have featured in research studies since the 1970s. The development of constructivism as a learning theory applicable to science has given rise to the use of constructivist-based strategies for teaching ideas regarded as 'hard' – meaning abstract or requiring a specific conceptual idea. These are well worth testing in the classroom – the (positive) effects on student learning can be dramatic and surprising. A teaching programme, CASE, has developed as a direct result of research applying Piaget's theory of learning to science. This has gained in popularity as a way of helping enhance GCSE Science results. Using CASE requires training and adherence to a set scheme. Practical work as a key tool in the science teacher's armoury has suffered from successive research studies indicating that its value in assisting science learning is much lower than teachers might automatically assume. This suggests that investing preparation time in devising activities that are 'active' but are alternatives to traditional experiments may be well spent. Context-based approaches to teaching science provide a wealth of such activities, but have not yet been thoroughly evaluated by researchers.

In general, perhaps, the most important message research on science education has for teachers is to think about their teaching entirely from the children's perspective. We normally regard the science curriculum or examination as the target. Teachers often say (and students often ask) 'Is this on the curriculum?' 'Will there be an examination question on this?', dismissing as 'time-wasting' topics, approaches or activities that do not fit into this tight regime. Research tells us to keep the curriculum in mind, but to focus on how children learn and on approaches that may be of greatest benefit to them. Teachers may benefit from considering these questions in planning the teaching of a topic:

- What would interest the children/students about this topic?
- What experiences might the children/students have had about this?
- What teaching strategy might make this 'come alive' for the class?
- What research findings directly relate to teaching this topic?
- What 'aha' moments may be possible/needed for understanding?
- How can I evaluate or assess the impact of what I do on the children's/students' learning?

Answering these will not at all detract from teaching the curriculum or preparing a class for an examination. If put into practice, students are more likely to experience positive teaching likely to hold their attention and encourage their learning. A teacher who can do this regularly and review his/her practice will be using research-based evidence in the best possible sense.

THINKING ABOUT PRACTICE

■ Consider your own experience of working in SCIENCE, either at university (research project/practical experiments) or in a job. What factors are characteristic of scientific research? Are these/should these be considered criteria for 'research' in any field? Explain.

■ Choose one of the four areas of research reviewed here. How might the findings from this area impact on your classroom practice?

■ How might you assess the impact on students' learning of activities you do in order to teach? Choose one topic from your teaching and propose a research question you could answer using one of the strategies proposed in Box 12.1 (pp. 226–7).

Chapter 13

Doing action research in science education

Nobody made a greater mistake than he who did nothing because he could only do a little.

(Edmund Burke)

INTRODUCTION: ACTION RESEARCH AS A FORM OF PROFESSIONAL DEVELOPMENT

In Chapter 11, we argued that critical reflection was essential for the professional science teacher. Being a professional implies commitment to professional development directed towards identified goals. These may focus on subject knowledge and/or pedagogy; about how science knowledge can best be transformed to teaching materials, guided by curriculum requirements to create a form of school science that is authentic, comprehensible and relevant to students. The conceptual difficulty of some ideas represented in SCIENCE makes these development needs inevitable; there are sure to be compromises in terms of the science, or students' understanding.

Teaching about a subject that is evolving, highly interlinked and often conceptually abstract, makes the teacher's job naturally challenging. Jerome Bruner suggested that any topic could be taught to a child of any age in some 'intellectually honest' way (1960). We like this notion that science teaching should be *intellectually honest*, but, in itself, this is a motto and does not show teachers *how* to do this.

SCIENCE is subject to political pressures and has economic, social and moral implications (see Chapter 3). The science curriculum is always going to be an area of contention. As science teachers we may wish to reflect SCIENCE, perhaps taking a view that science education, like SCIENCE, should be seen as morally neutral and that, like some scientists, we could argue that our concern is not the use made of it.

Supporting such an argument is difficult when a main justification for teaching science to all young people is to help them function in a technologically advanced democracy. Therefore, any science teacher evaluating their teaching (see Chapter 9) is likely to feel that there is room for improvement. Further, as we discussed in Chapter 11, the type of support needed may not always be readily available. Courses give insights into other teachers' thinking. However, the teacher is dealing with specific contexts – particular classes, with specific learners, in a unique and sometimes idiosyncratic institutional setting. Ideas from CPD courses and informal advice from colleagues still need to be tested with the teacher's own students.

Note the parallel here between teacher-learning and student-learning: just as the curriculum needs to be transformed into lesson plans to help students learn, so professional development inputs need to be translated into a form that provides a basis for action for the teacher. In both cases, the transformation process needs to take account of prior knowledge and classroom context.

Genuine educational research tends to be over-simplified and de-contextualised in many available courses, so we suggested in Chapter 11 that inductees will benefit if they can engage with the original literature where that is readily available. Chapter 12 reviewed some research evidence presented in the literature. Ultimately, there is only one way of finding out if what the literature presents will apply in a teacher's own setting. This is recognised in the notion of 'reader generalisability' that has been suggested for interpreting qualitative research (Kvale, 1996). This means that the research needs to be presented in sufficient detail to help the practitioner see whether the findings are likely to transfer to the teacher's own professional context.

We suggest that the final step in taking on the identity of a fully professional science teacher is to become a teacher-researcher in your own classroom. We argue that there is no better form of professional development than carrying out research into your own current practice, as a means to obtain the evidence to inform future practice.

WHAT IS ACTION RESEARCH?

We are suggesting participation in something that has become known as 'action research'. This term was suggested in the 1940s by the social psychologist Kurt Lewin who was concerned that research often derived from the researchers' theoretical considerations (as tends to be the case in SCIENCE, see Lakatos, 1978). He thought there was a need for research in the forms of programmes of action responding to identified social problems (Schwandt, 2001). Lewin developed a notion of a research cycle, with similarities to the way that teachers plan, teach and evaluate lessons and use the information gained to inform the next cycle of teaching. In educational circles, action research seems to have been taken up in two distinct ways. One of these involves academic researchers working with practitioners as partners in research. An example of this may be the *Children's Learning in Science Project*, CLiSP (see Box 13.1). The type of academic–practitioner relationship in this type of research has been discussed extensively, as there are real issues about the division of labour, responsibility and, most of all, *power* in the partnership (Zajano and Edelsberg, 1993; Moje, 2000).

A valid question might be, 'to what extent does the model used by CLiSP actually fit an action research pattern?' We suspect that obtaining a definitive answer to such a question is impossible, as there are different definitions and interpretations of action research. However, we believe that the *key issue* is the extent to which the practitioners feel they have *ownership* of the problem. Action research in teaching is a way of responding to perceived professional problems and issues as a means to improving professional practice. If the teachers in the CLiSP study shared the academic researchers' concerns (and we would admit that this can include being persuaded to take ownership of the problems) then this was action research for the practitioners. Otherwise, this is an example of research that derived from the theoretical concerns of the researchers, and in which they *then* involved teachers. This would be *collaborative* research, but is not really action research in the usual sense. This would not negate the potential value or validity of these studies, but it does make them something other than action research.

The second type of action research is research undertaken by practitioners, teachers themselves, to respond to an issue or problem that *they* have identified in their professional practice. In such a case, the label of action research is clearly less problematic, as the practitioners do have ownership of the research problem. This does not necessarily imply working alone. The practitioner setting out on action research may wish to involve departmental colleagues, LEA advisers, or perhaps academics from the local university. These colleagues may become partners in a full collaboration, or just advisers for the project.

BOX 13.1 ACTION RESEARCH IN CLiSP

The Children's Learning in Science Project (CLiSP) was based at the University of Leeds. The project divided into two phases. During the first phase data were collected in schools as part of a national survey, then analysed in a number of science topic areas such as particle theory (Brook *et al.*, 1984). Subsequently, the project moved into a curriculum development phase, producing 'constructivist' teaching schemes in three topic areas, in the light of the survey findings (Driver and Oldham, 1986). This later stage took on an action research perspective to the extent that local teachers were invited to work in collaboration with the university researchers, so that 'case-studies' with school classes could be undertaken. For example, for the topic 'particulate nature of matter' the research was carried out by a teacher-associate (a secondary teacher on secondment to CLiSP), in collaboration with 11 science teachers in schools, all of whom were involved in keeping diaries while teaching the topic and two of whom were observed teaching, and interviewed (Wightman, 1986). The school teachers were treated as co-researchers in a number of senses:

a   the group was provided with working documents for discussion and comment;
b   the teachers who were observed teaching for the case studies were allowed a veto of data collected;
c   the two observed teachers were cited as collaborating authors of the research report.

Following the collection of data on current practice the working group moved to a stage of writing, implementing and evaluating new curriculum materials based on a constructivist approach. The report on this stage of the work emphasises the action research nature of the exercise (Johnston and Driver, 1991) with the practising teachers intimately involved in the work, acting as co-workers with the university-based researchers.

The involvement of academic researchers does not debar the work from being action research *if* the initial agenda originates with the practitioners themselves, and the practitioners maintain control over the project. In research projects where teacher and academics (or other external researchers) work together, there is clearly the potential for a whole spectrum of types of partnerships, some of which will better fit the action research tag than others.

COLLABORATIVE RESEARCH ACTIVITIES AS PROFESSIONAL DEVELOPMENT

One reason why a teacher may wish, initially at least, to undertake classroom inquiry as part of a collaborative venture is a perceived lack of research skills. This is a reasonable concern for science teachers who seldom have a background that gives them a strong insight into the research methodologies used in the *social* sciences.

We believe that basic training in classroom research should be part of the PGCE qualification (see p. 254). At present, however, most newly qualified teachers feel limited confidence in their ability to undertake any formal kind of classroom research. In this context, working with local university researchers may be a very helpful form of professional development. Involvement in projects undertaken in the local university can be a way of making contact with local academics, finding out how to access useful seminars and other sources of information, and of starting to develop a greater awareness of how educational research occurs. Academics are usually keen to involve local teachers in their work – after all educational researchers need educational contexts in which to undertake their research.

When forming any kind of research relationship with a university department establishing ground rules at the start is a good idea. Sometimes academics are only interested in access to teachers, classrooms and students to carry out their existing research plans. In some cases, the university side may not explain much about what they are aiming to find out, as this could influence, for example, the teacher's behaviour in the classroom. Such research projects need access to schools and classrooms. We would encourage science teachers to be generous in providing that access subject to certain safeguards, for example, that you are convinced the research will not be detrimental to teaching and learning. However, involvement in this type of research does little for the professional development of the teacher concerned.

Of much more value is involvement in a project where teachers are offered a more active form of partnership. This may or may not include being part of the initial planning, but should certainly see the teachers as part of the research group, involving them in the on-going discussions about the progress of the work. In this way, the teacher begins to develop her own professional knowledge and skills.

Research partnerships and networks are becoming quite common in an attempt to involve more practitioners in research, to share good practice and disseminate research skills and findings. For example, one of the authors is involved as the university 'critical friend' to a partner school as part of a research partnership called 'SUPER' (see Box 13.2) between the Faculty of Education and schools. This is a genuine partnership in that the school partners are fully involved in decision-making and the strategic direction of the project.

THE NATURE OF ACTION RESEARCH

The type of involvement in research we suggest can be interesting, as well as a way of contributing to valuable work, but remember that research undertaken by academics has some features that are not shared in practitioner action research. Inevitably, the academic researcher is concerned with producing findings that are suitable for publication in some form, and that are therefore seen to have some relevance to educational contexts in general. An academic researcher who collects data in your science lesson is hoping that these may illuminate a research issue that has wider significance, so that other teachers may be able to find any conclusions that could have relevance for their practice.

There are important debates about the nature of generalisability in education research, especially in qualitative studies (Kvale, 1996; Taber, 2000b) that need not concern us here. However, research that is disseminated widely, needs to have followed certain accepted ways of working so that readers can make certain assumptions about how the data have been obtained and analysed. Usually, this means that the methodology is planned in advance, and the basic research plan needs to be followed in order to collect meaningful data that can be interpreted in ways that can inform the wider academic and professional communities. This is, of course, a principle common to the way that SCIENCE generally works.

Publication of action research undertaken by practitioners is not an imperative, so can be undertaken in a different way. The most common reason for a teacher undertaking action research is to improve professional practice, so the topic for research is situated in a particular 'problematised' context. As improving practice is the goal, planning and following precise research plans and protocols is less necessary. This

247

BOX 13.2 THE SCHOOLS–UNIVERSITY PARTNERSHIP FOR EDUCATIONAL RESEARCH

The 'SUPER' Project will address the national concern about the perceived lack of impact of the outcomes of educational research in the context of the Research Schools Partnership. It will do so by seeking to describe and document, analyse and interpret, conceptualise and understand, and report on:

- the evolving relationships between Partnership schools and between Partnership schools and [the University of Cambridge Faculty of Education];
- emerging practice-based research issues;
- practice-based research processes; their outcomes and effects.

It follows therefore that the Research Schools Partnership is the 'research site' for the 'SUPER' Project. The 'SUPER' Project aims to:

- observe, describe and document, analyse, interpret, conceptualise, understand and report on the processes, outcomes and effects of the practice-based research work undertaken within the evolving partnership;
- achieve these tasks by addressing an agenda of questions (a selection of which are listed below);
- concentrate, as a priority, on questions about understanding how a systemic research culture might be created within and across the Research Schools Partnership;
- compare issues that arise from within the Research Schools Partnership with those being reported from Partnership projects elsewhere;
- use developing understandings of the processes and outcomes of the work undertaken within the Research Schools Partnership to facilitate its constructive development;
- disseminate the outcomes of the practice-based research work of the Research Schools Partnership as widely as possible by appropriate means.

(From the SUPER website, www.educ.cam.ac.uk/rsp/ projects/super.html, accessed 12 April 2004) means that a teacher can stop trying something new as soon as s/he realises it does not seem to be working, rather than feeling the need to continue to collect enough data to write a convincing report.

Academic researchers and classroom practitioners need to plan their research, and collect data, and evaluate the evidence. However, the 'burden of proof' is different in these two situations. The academic researcher needs to be sure enough to justify recommending new practices to the profession. The evidence should convince beyond reasonable doubt. The classroom teacher is using research as a systematic way to find ways to do things better. The evidence collected allows the teacher to decide whether to change current practice on the balance of probabilities – on what seems to be the best course of personal action in the future. With the next cycle of planning teaching and evaluation the balance of the evidence may shift again, and practice move forward once more.

Support for the practitioner-researcher

Given that most science teachers lack training for carrying out classroom research, any practitioner wishing to undertake this type of research into their own practice must access support. We have suggested above that science education researchers in a local university may be supportive, but academic researchers must, by definition, focus on problems with a strong theoretical justification and the potential to produce publishable results. Universities are funded to undertake certain types of activities, and have to justify their existence through their teaching and commissioned or published research. We will return to university links in a moment. Some schools and colleges have limited research funds available internally, and at least one school known to the authors has a member of teaching staff designated as the 'school researcher' able to offer support and advice. Advanced Skills Teachers may also be able to assist in developing action research programmes.

Networks for educational action researchers – providing links, websites, journals and conferences aimed at the teacher practitioner (see Box 13.3) – are also available. The British Educational Research Association (BERA), although it includes many academics among its membership, is open to anyone involved in, or interested in, educational research (see Box 13.4).

Accessing support from academic researchers

Although education faculty in universities are not funded to freely offer research support to teachers on request, there are ways that the classroom practitioner can access advice and expertise from the local university. Often universities will allow teachers in local schools access to libraries.

BOX 13.3 THE COLLABORATIVE ACTION RESEARCH NETWORK

The quality of our work in the professions depends upon our willingness to ask questions of ourselves and others, and to explore challenging ideas and practices, including the values that underpin them. The Collaborative Action Research Network (CARN) is committed to supporting and improving the quality of professional practice, through systematic, critical, creative inquiry into the goals, processes and contexts of professional work . . .

CARN aims to encourage and support action research projects (personal, local, national and international), accessible accounts of action research projects, and contributions to the theory and methodology of action research.

CARN sets out to promote recognition:

- that professional development requires critical inquiry into past, current and future practice;
- that practitioners themselves should be actively and creatively involved in defining and developing professional practice;
- that practitioners themselves should contribute to the growth of valid professional knowledge and theory;
- that all relevant communities (including service-users, students, clients, etc.) need to be involved in developing the provision of services;
- that individual professional development needs to be seen in the context of institutional practices and structures;
- that action research provides a powerful means of developing worthwhile professional and institutional practice . . .

(From the CARN website,
www.did.stu.mmu.ac.uk/carn/, accessed April 12 2004)

BOX 13.4 THE BRITISH EDUCATIONAL RESEARCH ASSOCIATION

The aim of the Association is to sustain and promote a vital research culture in education by encouraging:

- an active community of educational researchers by promoting co-operation and discussion:
  - with policy makers, institutional managers and funding agencies;
  - with other national educational research associations, international associations and the European Educational Research Association;
  - with other researchers in the social sciences and related areas of work; and
  - with teachers and lecturers and their associations

by encouraging and supporting:

- debate about the quality, purpose, content and methodologies of educational research

by developing and defending:

- an independent research culture committed to open inquiry and the improvement of education

by enhancing:

- the professional service it provides for its members;
- effective communication and discussion within BERA; and
- the training and education of educational researchers, their effectiveness, conditions of work and rights.

(From the BERA website,
www.bera.ac.uk/, accessed 12 April 2004)

Many schools already have contacts with academics through PGCE partnerships. Where such a relationship already exists, then university tutors may see part of their roles as providing support to science teachers in departments mentoring trainees. In general, though, obtaining support from the local university will normally involve one of the following:

- becoming involved in an area of research in which academics are already active;
- forming a formal mentoring arrangement;
- enrolling on an official course.

Bearing in mind our comments above, the first option can only help the teacher undertake action research if the university research interest overlaps with the problem area identified by the practitioner. For example, one of the authors is involved in a collaborative project known as APECS (Able Pupils Experiencing Challenging Science – see Box 13.5). As part of this project, a series of seminars for interested parties was arranged. Some teachers attending the seminars had a particular professional concern about the extent to which the most able students were being 'stretched' in their science lessons and attended to discuss the issues and pick up ideas to try out. Some reported their own experiences of trying out new ideas to subsequent seminar meetings.

Formal research mentoring of teachers became relatively common under the BPRS scheme – the Best Practice Research Scholarships. This was a government-funded initiative to support teacher research, which

---

BOX 13.5 ABLE PUPILS EXPERIENCING CHALLENGING SCIENCE

The APECS Project is a collaborative research project between the Universities of Cambridge, Reading and Roehampton. The project is intended to explore the issue of providing suitably challenging experiences for the most able pupils in science lessons. The project is currently supported by the University of Cambridge Faculty of Education through its research seminar series: 'Meeting the Needs of the Most Able in Science.' This seminar series is providing a forum for the research team to meet with interested practitioners – to discuss the issue of meeting the needs of the most able pupils in science classes, and to plan and facilitate classroom-based teacher-research. An e-mail newslist has been established for anyone wishing to be kept informed about the project (contact: APECSproject-owner@yahoogroups.com).

(From the APECS website:
www.educ.cam.ac.uk/apecs/ accessed 12 April 2004)

---

allowed scholarship holders to use part of the award to buy support from a suitable mentor with expertise to support classroom research (such as a university academic). Although the BPRS programme has now finished, it has been announced that similar levels of funding will instead be channelled directly to schools, so in principle such mentoring arrangements can continue.

Research training for teachers?

Our last suggestion is that teachers consider undertaking a higher degree in education with a strong research component. In SCIENCE, as we discussed in Chapter 2, becoming a fully fledged scientist normally involves a period of research training while working for a higher degree. Obtaining a PhD in a science discipline means being recognised as someone who *does* science – rather than someone who *knows about* science. SCIENCE is above all else a set of research traditions. The *approaches to* research in education and natural science may vary (see Chapter 11), but we would expect science teachers to appreciate that the ultimate professional practitioner is one who contributes to the field through research. Ideally, science teachers should follow a development path that leads to a formal induction into educational research. After first degree and teacher training, and completion of formal induction, there should be the opportunity for further post-graduate level study, to include training in using the research literature (cf. Chapter 12), educational research methods, and a supervised research project. Whether at masters or doctoral level this would enable science teachers to complete their professional training, and the transition to a fully professional science teacher: well prepared to plan, execute, evaluate and research their professional practice of science teaching.

RESEARCH TRAINING AS PART OF INITIAL TEACHER EDUCATION

We clearly see that being a *fully* professional science teacher means seeing teaching not only as an evidence-based activity, but also to some extent as a research-based activity (Taber, 2000c). The idealised professional science teacher identifies problems and issues that need addressing, and then searches for relevant evidence – initially from the literature, but then by empirical investigation of classroom practice.

253

We therefore believe that, as part of teacher education, all inductees should be empowered with the basic knowledge, understanding and skills to research their own professional practice. Sadly, in practice, this is not currently always the case. Universities often set PGCE assignments that involve some limited forms of classroom enquiry, but in view of the need to 'cover' all the mandatory QTS criteria, there is very little time in a PGCE course to provide substantive research training. It may sometimes be possible to offer trainees opportunities to get involved in research projects, but this is likely to be on an optional elective basis. For example, a group of trainee science teachers on the partnership PGCE course at the University of Cambridge volunteered to take part in classroom-based research looking at teaching about the Ideas and Evidence strand of Sc1 at KS3 for the KS3 strategy (see Box 13.6) – an aspect of the science curriculum known to be problematic (Driver *et al.*, 1996).

The PGCE is a post-graduate qualification, required to be substantially at the level of a Master's degree (see pp. 255–6). Before the introduction of the QTS standards, the PGCE had long been established as a respected post-graduate university qualification. In more recent years, as the QTS standards have been so central in initial teacher education, there has been the danger that PGCE courses can be 'hijacked', and dominated, by considerations of how the courses should provide trainees opportunities to demonstrate they are meeting the standards. Two-thirds of the time on a PGCE course is in schools and the QTS standards provide a framework for planning and assessing trainees' work on placements. However, 'the standards' can tend to dominate thinking during the third of the course spent in the university as well.

We are certainly *not* suggesting that there should be a sharp division between 'academic' and 'practical' concerns on the PGCE. Rather, we see professional practice as a kind of on-going dialogue between theory and action, between doing and thinking. Teaching has an iterative nature, with cycles of planning (see Chapters 6 and 7), teaching (Chapter 8) and evaluation (Chapter 9). For this to be effective, *initial teacher education must provide a repertoire of relevant theoretical frameworks for conceptualising, analysing and reflecting on classroom observations and actions.* This aspect of training is in danger of being marginalised if there is too much concern that all PGCE course inputs are QTS standards-related.

We see a parallel here with what is happening in school-level education itself: the danger of 'teaching-to-the-test'. Universities, just like

BOX 13.6 KS3 STRATEGY 'SCIENCE ENRICHMENT PROJECT' – ON IDEAS AND EVIDENCE

University Education Faculties were invited by the National Key Stage 3 Strategy to undertake a 'Science Enrichment Project' to 'enrich existing initial teacher education and training about the "Ideas and Evidence" in science aspect of Scientific Enquiry (Sc1).' . . . The successful Cambridge project bid focused on the related themes of the roles of theories, models and explanations in science and the development of science, and was linked to an existing series of seminars exploring the theme of *Meeting the Needs of the Most Able in Science.*

The Cambridge project was based around the work of trainee teachers working with KS3 (Y7 and/or Y8 and/or Y9) classes in partner schools. The trainees developed, as part of their planning, teaching and evaluation of lessons, materials and activities that:

■ helped students learn about aspects of science that relate to the 'ideas and evidence' aspect of Sc1 in the National Curriculum;
■ were suitably differentiated to meet the needs of the most able students in KS3 groups.

The outcomes of the project, along with those from other partner universities (Keele, King's College London, London Institute of Education, and York), are being disseminated to teachers through support from the Gatsby charitable foundation's Science Enhancement Programme.

---

schools, can fall into the trap of focusing on what will be assessed – and the QTS standards are likely to be the focus of Ofsted inspections visits. While the QTS standards are useful, a PGCE is meant to demonstrate a critical intellectual engagement with education as a field of study. The foundation subjects on which Educational Studies has its intellectual foundation (i.e. the philosophy, history, sociology and psychology of education) have become subdued, and perhaps sometimes almost invisible, compared with the teaching standards.

The national framework for qualifications requires a post-graduate certificate course to be at Master's (M) level: 'A programme leading to a Graduate Certificate or Graduate Diploma might have some M level outcomes, but use of the Postgraduate title for the award would be justified only if most or all of the outcomes were assessed at M level' (QAA, 2001). This should provide some protection for academics in education faculties trying to ensure that the PGCE retains its academic standards, and does not become reduced to little more than a course that is preparing trainees to meet QTS. We note with interest and a little concern, that there has been some discussion of changing the PGCE to make it a graduate rather than post-graduate level qualification (UCET, 2003), and some Higher Education Institutions are already offering their PGCE as a *Professional Graduate Certificate in Education*.

As well as ensuring that key ideas from the foundations of educational studies are included (in some form) in PGCE courses, another key way to maintain PGCE standards is the inclusion of research projects involving the collection and analysis of classroom data, linked to theoretical concerns from the literature (cf. Chapter 12). Perhaps opportunities to be involved in 'research teams' (such as on the 'Ideas and Evidence' project, Box 13.6) could become a regular part of initial teacher education. Some of the more substantial PGCE assignments set on some courses have something of an action research flavour, although with limited support in planning, executing and analysing *social science research*. Including modest training for, and experience in, carrying out small-scale classroom-based research is an ideal way of maintaining the academic status of the PGCE. Allowing that aspect of the PGCE course to be credited towards Masters' degrees, where further research training and experience is available, would emphasise a professional development route giving progression towards the status of teacher-researcher – perhaps the ultimate stage of development for the professional science teacher.

THINKING ABOUT PRACTICE

- What are the advantages to a science teacher of participating in action research?
- Choose one of the projects mentioned in the chapter. Visit the website and prepare a short review of the aims, objectives, methodologies and findings of the project. How might the findings impact on your practice?

Coda

Moving science teachers and school science forward

We hope you have enjoyed reading this book, but we also hope it has made you think. Perhaps you agree with much of what we have written, but if not, then at least we hope your disagreement is based on genuine engagement with the issues – with critical reflection upon what you have read.

We started the book by suggesting that:

science teachers are not scientists;
school science is not SCIENCE.

This was not an attempt to be provocative for its own sake but, rather, to establish a position from where the inductee's professional development can move forward. If you have read this far, then we hope you appreciate these two propositions. We do not ask you to either agree or disagree with either claim; rather, we believe they are both true in a sense, an important sense, and we very much hope that *you understand the sense in which these statements are true.*

School science is not just a subset or watered down version of SCIENCE, but a socially constructed curriculum subject designed to meet the requirements and preferences of certain pressure groups – professional bodies, universities, political parties, employers' groups and so on. A professional science teacher's role includes transforming curriculum science into effective, enjoyable, worthwhile lessons. Another aspect of the science teacher's role is to be part of the larger profession and help make sure the profession exerts its influence on curriculum and assessment regimes. Your job is to transform the paper curriculum into educational activity in the classroom, and to help develop something of greater educational worth at the level of official policy and documentation.

Science teachers are not scientists who happen to be working in education. Their primary professional identity is as *teachers*, working with young people to help them understand key aspects of science and appreciate why they should value such understanding. The science teacher's area of special expertise is much wider, if not as deep as that of a professional scientist. The science teacher will not necessarily understand any science area as well as the research scientist, but will understand how to transform and communicate science with considerable expertise. The science teacher applies three areas of knowledge – science, pedagogy and the students – to construct science knowledge appropriately. In preparing to teach, the science teacher comes to understand much science in more depth than most scientists, who are often specialists in a limited, specific area. If, as we recommend, the science teacher conceptualises science as comprised of socially constructed models to be reflected and reconstructed through classroom discourse, then s/he is likely to develop a much better understanding of science per se than most professional scientists.

Professional scientists can claim that they are pushing back the frontiers – actually *doing* SCIENCE, creating new knowledge through their research. Few science teachers have time or facilities to undertake research in a science discipline. When done well, teaching – like SCIENCE – is an evidence-based profession, drawing on well-developed theoretical frameworks and collecting, analysing and interpreting relevant data. The constant cycle of prepare–teach–evaluate is a sequence of real-life ongoing 'educational experiments'. The effective science teacher, like the effective scientist, identifies problems, forms hypotheses about their causes and potential solutions, and then sets out on research. The science teacher's action research usually produces knowledge that is largely of local value and of limited reliability, but contributes to on-going critical reflection and improvement in professional practice.

We have tried to create an image of the 'professional science teacher' that both reflects the many professionals we work with and provides a target for inductees to aspire to. Some of the points that we make in this book, relating to principles of lesson planning for example, could apply to new teachers whatever their teaching subject. SCIENCE, though, is a key cultural activity with its own place in society and science teachers are more than just teachers who happen to know some science. If school science is to justify its special place in the core curriculum, then it needs to reflect something of the special nature of SCIENCE – hence, science teachers need to be more than teachers who teach school science. Science teachers are the people who can best appreciate how the special attributes of SCIENCE (perhaps the values we refer to in Chapter 2) can contribute to the education of young people.

The profession of science teacher is noble, exciting and socially valuable. Remember, the science curriculum is socially constructed, and it is up to you what you make of it, for, and with, your students. As a science teacher, reflect your personal and professional values in all that you do in your professional role. If something is not working well, *you* have the potential to find out why and work to make it better. Most of all, you have the opportunity to present to young people an image of SCIENCE that you find important, engaging and educationally valuable *for them*.

Construct SCIENCE knowledge with your classes, and enjoy the process.

SUGGESTIONS FOR FURTHER CRITICAL READING

In the introduction to this book we suggested:

> If you are a new entrant to science teaching, then we would encourage you to read *this* book first and use the ideas presented here to inform your subsequent (critical and reflective) reading.

There are many useful books about science education worth reading. We would recommend that it is better to take time to critically engage with texts than to try to read a great deal. The books in the following selection have something in common: they are each edited collections with contributions from a range of authors, and so include a wide range of perspectives. (We leave it for the reader to decide to what extent they reflect a common position in terms of the authors' basic values.) Some of the chapters in some of these books are slightly out of date in terms of the technical changes in curriculum and examination specifications, but the issues raised will remain important and central to science education for a long time.

Amos, S. and Boohan, R. (eds) (2002) *Teaching Science in Secondary Schools: perspectives on practice*, London: RoutledgeFalmer.

Amos, S. and Boohan, R. (eds) (2002) *Teaching Science in Secondary Schools: a reader*, London: RoutledgeFalmer.

Cross, Roger (ed.) (2003) *A Vision for Science Education: responding to the work of Peter Fensham*, London: RoutledgeFalmer.

Levinson, Ralph (ed.) (1994) *Teaching Science*, London: Routledge.

Millar, R., Leach, J. and Osborne, J. (eds) (2000) *Improving Science Education: the contribution of research*, Buckingham: Open University Press.

Sears, J. and Sorenson, P. (eds) (2000) *Issues in Science Teaching*, London: RoutledgeFalmer.

Wallace, J. and Louden, W. (eds) (2002) *Dilemmas of Science Teaching: perspectives on problems and practice*, London: Routledge.

References

Adey, P. and Shayer, M. (1993) An exploration of long-term far-transfer effects following an extended intervention programme in the high school science curriculum, *Cognition and Instruction*, 2 (1), pp. 190–220.

Adey, P. and Shayer, M. (1994) *Really Raising Standards: cognitive intervention and academic acheivement*, London: Routledge.

Adey, P., Shayer, M. and Yates, C. (1989) Thinking Science: The Curriculum Materials of the CASE Project, Nelson Thornes.

Aikenhead, G. S. (2003) Chemistry and physics instruction: integration, ideologies, and choices, *Chemistry Education: Research and Practice in Europe*, 4 (2), pp. 115–30.

Al-Kunifed, Ali and Wandersee, James H. (1990) One hundred references to concept mapping, *Journal of Research in Science Teaching*, 27 (10), pp. 1069–75.

American Chemical Society (1994) *Chemistry in Context*, Dubuque, Iowa: Kendall-Hunt.

Amos, S. and Boohan, R. (eds) (2002a) *Aspects of Teaching Secondary Science: perspectives on practice*, London: Routledge Falmer.

Amos, S. and Boohan, R. (eds) (2002b) *Teaching Science in Secondary Schools: a reader*, London: Routledge Falmer.

Anderson, J. R. (1995) *Learning and Memory: an integrated approach*, New York: John Wiley & Sons.

Anderson, Lorin W. and Krathwohl, David R. (2001) *A Taxonomy for Learning, Teaching and Assessing: a revision of Bloom's taxonomy of educational objectives*, New York: Longman.

Assessment of Performance Unit (APU) (1988a) *Science at Age 11: a review of APU survey findings 1980–1984*, London: HMSO.

Assessment of Performance Unit (APU) (1988b) *Science at Age 15: a review of APU survey findings 1980–1984*, London: HMSO.

Assessment of Performance Unit (APU) (1989a) *National Assessment: the APU science approach*, London: HMSO.

Assessment of Performance Unit (APU) (1989b) *Science at Age 13: a review of APU survey findings 1980–1984*, London: HMSO.

Association for Science Education (ASE) (1978) *Science for Less Academically Motivated Pupils* (LAMP project), Hatfield: ASE.

Association for Science Education (ASE) (1979) *Alternatives for Science Education*, Hatfield: ASE.

Association for Science Education (ASE) (1986) *Science and Technology in Society (SATIS)*, Hatfield: ASE.

Ausubel, David P. (1961) In defense of verbal learning, *Educational Theory*, 11, pp. 15–25.

Ausubel, David P. (1968) *Educational Psychology: a cognitive view*, New York: Holt, Rinehart & Winston.

Ausubel, David P. (2000) *The Acquisition and Retention of Knowledge: a cognitive view*, Dordrecht: Kluwer Academic Publishers.

Ausubel, David P. and Robinson, Floyd G. (1971 [1969]) *School Learning: an introduction to educational psychology*, London: Holt International Edition (first published by Holt, Rinehart & Winston, 1969).

BAAS (1918) *British Association for the Advancement of Science – Report 1917*, London: Murray.

Bachelard, Gaston (1968) *The Philosophy of No: a philosophy of the scientific mind*, New York: Orion Press (original French edition published in 1940).

Banks, P. (1997) Students' understanding of chemical equilibrium. Unpublished MA thesis, University of York.

Barker, V. (1994) A longitudinal study of 16–18-year-old students' understanding of basic chemical ideas. Unpublished D.Phil. thesis, University of York.

Barker, V. and Millar, R. (2000) Students' reasoning about basic chemical thermodynamics and chemical bonding: what changes occur during a context-based post-16 chemistry course? *International Journal of Science Education*, 22 (11), pp. 1171–200.

Barker, V., Sang, D. and Shorter, A. (2002) *Building Success in GCSE Science: chemistry, physics and biology*, Dunstable: Folens.

Bennett, J. (2003) *Teaching and Learning Science: a guide to recent research and its applications*, London: Continuum.

Black, P. (1993) The purposes of science education. In Richard Hull (ed.) *ASE Science Teachers' Handbook*, Hemel Hempstead: Simon & Schuster.

Black, P. and Wiliam, D. (1998) *Inside the Black Box: raising standards through classroom assessment*, available at www.pdkintl.org/kappan/kbla9810.htm, accessed 1 August 2004.

Bloom, B. S. (1964) The cognitive domain. Reprinted in L. H. Clark (1968) *Strategies and Tactics in Secondary School Teaching: a book of readings*, London: Macmillan, pp. 49–55.

Bodner, George M. (1986) Constructivism: a theory of knowledge, *Journal of Chemical Education*, 63 (10), pp. 873–78.

Brook, A., Briggs, H. and Driver, R. (1984) *Aspects of Secondary Students' Understanding of the Particulate Nature of Matter*, Children's Learning in Science Project, Leeds: Centre for Studies in Science and Mathematics Education, University of Leeds.

Bruner, Jerome S. (1960) *The Process of Education*, New York: Vintage Books.

Burton, G., Holman, J., Pilling, G. and Waddington, D. (1994) *Salters Advanced Chemistry: storylines, chemical ideas*, Oxford: Heinemann.

Cathcart, Brian (2004) *The Fly in the Cathedral*, London: Penguin.

Charmaz, Kathy (1995) Grounded theory. In Jonathan A. Smith, Rom Harrè and Luk Van Langenhove (eds) *Rethinking Methods in Psychology*, London: Sage Publications, pp. 27–49.

Children's Learning in Science Project (CLiSP) (1984, 1985) Full reports, University of Leeds: Centre for Studies in Science and Mathematics Education.

Claxton, G. (1990) *Teaching to Learn: a direction for education*, London: Cassell.

Cohen, L., Manion, L. and Morrison, K. (2000) *Research Methods in Education* (5th edn), London: RoutledgeFalmer.

de Jong, Onno (2000) Crossing the borders: chemical education research and teaching practice, *University Chemistry Education*, 4 (1), pp. 29–32.

DES (1979) Aspects of Secondary Education in England: A Survey by HM Inspectors of Schools London: HMSO.

DES (1989) Circular no.6/89 The Education Reform Act 1988: National Curriculum: Mathematics and Science under Section 4, London: HMSO.

DfEE/QCA (1999) *Science: The National Curriculum for England*, London: Department for Education and Employment/Qualifications and Curriculum Authority.

DfES (2002a) *Framework for Teaching Science: Years 7, 8 and 9*, Key Stage 3 National Strategy, London: Department for Education and Skills.

DfES (2002b) *Planning and Implementing Progression for Science in the Classroom* (Resource pack), Key Stage 3 National Strategy, London: Department for Education and Skills.

DfES (2003a) *Subject Specialism; Consultation Document*, London: Department for Education and Skills.

DfES (2003b) *Strengthening Teaching and Learning of Energy in Key Stage 3 Science* (Support Pack), Key Stage 3 National Strategy, London: Department for Education and Skills.

DfES (2003c) 14–19: Opportunity and Excellence, London: Department for Education and Skills.

Doherty, Anne, Goodwin, Alan and Benson, Ann (2000) *Trends in Recruitment of Secondary Chemistry Teachers onto 1-year PGCE courses*, Royal Society of Chemistry Chemical Education Research Group, available at www.rsc.org/lap/rsccom/dab/educ002activities.htm, accessed 1 August 2004.

Donaldson, M. (1978) *Children's Minds*, London: Fontana.

Driver, Rosalind (1983) *The Pupil as Scientist?*, Milton Keynes: Open University Press.

Driver, Rosalind and Bell, B. (1986) Students' thinking and the learning of science: a constructivist view, *School Science Review*, 67 (240), pp. 443–56.

Driver, Rosalind and Millar, Robin (1986) *Energy Matters – Proceedings of an Invited Conference: teaching about energy within the secondary science curriculum*, Leeds: Centre for Studies in Science and Mathematics Education.

Driver, Rosalind and Oldham, Valerie (1986) A constructivist approach to curriculum development in science, *Studies in Science Education*, 13, pp. 105–22.

Driver, Rosalind, Leach, John, Millar, Robin and Scott, Phil (1996) *Young People's Images of Science*, Buckingham: Open University Press.

Driver, Rosalind, Squires, Ann, Rushworth, Peter and Wood-Robinson, Valerie (1994) *Making Sense of Secondary Science: research into children's ideas*, London: Routledge.

Edwards, Derek and Mercer, Neil (1987) *Common Knowledge: the development of understanding in the classroom*, London: Routledge.

Eijkelhof, H. and Kortland, K. (1988) Broadening the aims of physics education. In P. Fensham (ed.) *Development and Dilemmas in Science Education*, Lewes: Falmer Press.

Ellis, H. C. and Hunt, R. R. (1989) *Fundamentals of Human Memory and Cognition* (4th edn), Dubuque, Iowa: William C. Brown Publishers.

Fensham, Peter J., Gunstone, R. F., and White, R. T. (eds) (1994) *The Content of Science: a constructivist approach to its teaching and learning*, London: Falmer Press, pp. 147–60.

Feyerabend, P. (1988) *Against Method* (rev. edn), London: Verso.

Furlong, J., Barton, L., Miles, S. *et al.* (2000) *Teacher Training in Transition: Reforming Professionalism*, Buckingham: Open University Press.

Gardner, Howard (1993) *Frames of Mind: the theory of multiple intelligences* (2nd edn), London: Fontana.

Gardner, Howard (1999) *Intelligence Reframed: multiple intelligences for the twenty first century*, New York: Basic Books.

Gardner, P. and Gauld, C. (1990) Labwork and students' attitudes. In E. Hegarty-Hazel (ed.) *The Student Laboratory and the Science Curriculum*, London: Routledge.

Gilbert, J. K. and Boulter, C. J. (2000) *Developing Models in Science Education*, Dordrecht: Kluwer Academic Publishers.

Gilbert, John K. and Zylbersztajn, Arden (1985) A conceptual framework for science education: the case study of force and movement, *European Journal of Science Education*, 7 (2), pp. 107–20.

Gilbert, John K., Osborne, Roger J. and Fensham, Peter J. (1982) Children's science and its consequences for teaching, *Science Education*, 66 (4), pp. 623–33.

Glaser, Barney G. (1978) *Theoretical Sensitivity: advances in the methodology of grounded theory*, California: The Sociology Press.

Glaser, Barney G. and Strauss, Anselm L. (1967) *The Discovery of Grounded Theory: strategies for qualitative research*, New York: Aldine de Gruyter.

Goodwin, A. (2001) Teachers' continuing learning of chemistry: implications for pedagogy. The Royal Society of Chemistry Chemical Education Research Group Lecture 2001, available at www.rsc.org/lap/rsccom/dab/educ002 activities.htm, accessed 1 August 2004.

Gott, R. and Duggan, S. (1995) *Investigative Work in the Science Curriculum*, Buckingham: Open University Press.

Hargreaves, Linda and Galton, Maurice (2002) *Transfer from the Primary Classroom – 20 years on*, London: RoutledgeFalmer.

Herrington, Neil and Doyle, Lesley (1997) *Curriculum Continuity between Primary and Secondary School*, London: Teacher Training Agency.

Hodson, D. (1990) A critical look: at practical work in school science, *School Science Review*, 71 (256), pp. 33–40.

ILEA (1987) *Science in Process*, Ten Units and Teachers' Guide, London: Heinemann.

Inhelder, B. and Piaget, J. (1958) *The Growth of Logical Thinking from Childhood to Adolescence*, New York: Basic Books.

Jenkins, E. W. (1989) Processes in science education: a historical perspective. In J. Wellington (ed.) *Skills and Processes in Science Education: a critical analysis*, London: Routledge.

Johnston, Kate and Driver, Rosalind (1991) *A Case Study of Teaching and Learning about Particle Theory: a constructivist teaching scheme in action*, Children's Learning in Science Project, Leeds: Centre for Studies in Science and Mathematics Education, University of Leeds.

Johnstone, A. H. (1989) Some messages for teachers and examiners: an information processing model, *Research in Assessment VII: Assessment of Chemistry in Schools*, London: Royal Society of Chemistry Education Division, pp. 23–39.

Jones, M. and Gott, R. (1998) Cognitive acceleration through science education: alternative perspectives, *International Journal of Science Education*, 20 (7), pp. 755–68.

Justi, Rosária and Gilbert, John (2000) History and philosophy of science through models: some challenges in the case of 'the atom', *International Journal of Science Education*, 22 (9), pp. 993–1009.

Kelly, G. (1955) *The Psychology of Personal Constructs*, New York: Norton.

Koestler, Arthur (1979) *Janus: a summing up*, London: Pan Books (first published by Hutchinson & Co., 1978).

Krathwohl, David R. (2002) A revision of Bloom's Taxonomy: an overview, *Theory Into Practice*, Autumn, accessed via www.findarticles.com, 24 February 2004.

Kuhn, Thomas S. (1977) *The Essential Tension: selected studies in scientific tradition and change*, Chicago: University of Chicago Press.

Kuhn, Thomas S. (1996) *The Structure of Scientific Revolutions* (3rd edn), Chicago: University of Chicago. (First edition published in 1962.)

Kvale, Steinar (1996) *InterViews: an introduction to qualitative research interviewing*, Thousand Oaks, California: SAGE Publications.

Kyriacou, C. (1998) *Essential Teaching Skills* (2nd edn), Cheltenham: Stanley Thornes.

Lakatos, Imre (1978) *The Methodology of Scientific Research Programmes*, Philosophical Papers, Volume 1, (edited by John Worrall and Gregory Currie), Cambridge: Cambridge University Press, 1978.

Lemke, Jay L. (1990) *Talking Science: language, learning, and values*, Norwood, NJ: Ablex Publishing Corporation.

Leo, E. and Galloway, D. (1995) Conceptual links between cognitive acceleration through science education and motivational style: a critique of Shayer and Adey, *International Journal of Science Education*, 18 (1), pp. 35–49.

LSRC (2004) Should we be using learning styles? What research has to say to practice, Learning and Skills Research Centre.

265

McCloskey, Michael (1983) Intuitive Physics, *Scientific American*, 248 (4), pp. 114–22.

Millar, Robin (1989a) Constructive criticisms, *International Journal of Science Education*, 11 (special issue), 1989, pp. 587–96.

Millar, Robin (1989b) What is 'scientific method' and can it be taught? In J. Wellington (ed.) *Skills and Processes in Science Education: a critical analysis*, London: Routledge.

Millar, Robin (ed.) (1989c) *Doing Science: images of science in science education*, London: The Falmer Press.

Millar, Robin and Osborne, J. (1998) *Beyond 2000: science education for the future*, London: King's College.

Miller, George A. (1968) The magical number seven, plus or minus two: some limits on our capacity for processing information. In *The Psychology of Communication: seven essays*, Harmondsworth: Penguin, pp. 21–50.

Moje, Elizabeth B. (2000) Changing our minds, changing our bodies: power as embodied in research relations, *International Journal of Qualitative Studies in Education*, 13 (1) pp. 25–42.

Moore, A. (2000) *Teaching and Learning: pedagogy, curriculum and culture*, London: RoutledgeFalmer.

Mortimer, E. F. and Scott, P. H. (2003) *Meaning Making in Secondary Science Classrooms*, Maidenhead: OUP/McGraw-Hill.

NACCCE (1999) *All our Futures: creativity, culture and education*, National Advisory Committee on Creative and Cultural Education, London: DfEE.

New Zealand Ministry of Education (1993) *Science in the New Zealand Curriculum*, Wellington: Learning Media.

Novak, J. (1978) An alternative to Piagetian psychology for science and mathematics education, *Studies in Science Education*, 5, pp. 1–30.

Novak, Joseph D. (1990) Concept mapping: a useful tool for science education, *Journal of Research in Science Teaching*, 27 (10), pp. 937–49.

Nuffield 11–13 (1986a) *How Scientists Work*, Teachers' Guide 1, London: Longman.

Nuffield 11–13 (1986b) *How Science is Used*, Teachers' Guides 2, London: Longman.

Ogborn, Jon, Kress, Gunther, Martins, Isabel and McGillicuddy, Kieran (1996) *Explaining Science in the Classroom*, Buckingham: Open University Press.

Osborne, J. and Collins, S. (2000) *Pupils' and Parents' Views of the School Science Curriculum*, London: King's College.

Osborne, J., Driver, R. and Simon, S. (1996) *Attitudes to Science: a review of research and proposals for studies to inform policy relating to uptake of science*, London: King's College.

Osborne, R. and Freyberg, P. (1985) *Learning in Science: the implications of children's science*, Auckland: Heinemann.

Parkinson, John (2004) *Improving Secondary Science Teaching*, London: RoutledgeFalmer.

Pollard, Andrew (2002) *Reflective Teaching: effective and evidence-informed professional practice*, London: Continuum.

Pope, M. L. (1982) Personal construction of formal knowledge, *Interchange*, 13 (4), pp. 3–14.

Pope, Maureen and Watts, Mike (1988) Constructivist goggles: implications for process in teaching and learning physics, *European Journal of Physics*, 9, pp. 101–09.

Popper, K. (1959) *The Logic of Scientific Discovery*, London: Hutchinson.

Popper, Karl R. (1979) *Objective Knowledge: an evolutionary approach* (rev. edn), Oxford: Oxford University Press.

Posner, G., Strike, K., Hewson, P. and Gertzog, W. (1982) Accommodation of a scientific conception: towards a theory of conceptual change, *Science Education*, 66 (2), pp. 211–27.

QAA (2001) *The Framework for Higher Education Qualifications in England, Wales and Northern Ireland*, The Quality Assurance Agency for Higher Education, January 2001, available at www.qaa.ac.uk/, accessed 8 May 2004.

Ramsden, J. (1992) If it's enjoyable is it science?, *School Science Review*, 73 (265), pp. 65–71.

Ramsden, J. (1997) How does a context–based approach influence understanding of key chemical ideas at 16+?, *International Journal of Science Education*, 19 (6), pp. 697–710.

Reed, Tomos (2004) To what extent can assessment for learning contribute to year 10 students' ability to meet the coursework requirements of GCSE Twenty First Century Science and promote scientific literacy? Unpublished PGCE Study-in-Depth, University of Cambridge.

Reiss, M. (2002a) What is science? In S. Amos and R. Boohan (eds) *Teaching Science in Secondary Schools: a reader*, London: RoutledgeFalmer.

Reiss, M. (2002b) Science education for all. In S. Amos and R. Boohan (eds) *Aspects of Teaching Secondary Science*, London: RoutledgeFalmer.

Riding, Richard and Rayner, Stephen (1998) *Cognitive Styles and Learning Strategies*, London: David Fulton.

Ross, K., Lakin, L. and Callaghan, P. (2000) *Teaching Secondary Science: constructing meaning and developing understanding*, London: David Fulton.

Rudduck, Jean and Flutter, Julia (2000) Pupil participation and pupil perspective: 'carving a new order of experience', *Cambridge Journal of Education*, 30 (1), pp. 75–89.

Schwandt, T. A. (2001) *Dictionary of Qualitative Inquiry* (2nd edn), Thousand Oaks, CA: Sage Publications.

SCISP (1974) *Schools Council Integrated Science Project*, Schools Council.

Scott, Philip (1987) *A Constructivist View of Learning and Teaching in Science*, Leeds: Centre for Studies in Science and Mathematics Education – Children's learning in science project.

Scott, Philip (1998) Teacher talk and meaning making in science classrooms: a review of studies from a Vygotskian perspective, *Studies in Science Education*, 32, pp. 45–80.

Screen, P. (1986) *Warwick Process Science*, Southampton: Ashford Press.

Shayer, M. and Adey, P. (1981) *Towards a Science of Science Teaching: cognitive development and curriculum demand*, Oxford: Heinemann Educational Books.

Smith, J. (1994) Are students' attitudes to science indicative of their general attitude to school and is there a gender bias in such attitudes? Unpublished MS thesis, University of York.

Solomon, J. (1987) Social influences on the construction of pupils' understanding of science, *Studies in Science Education*, 14, pp. 63–82.

Solomon, Joan (1992) *Getting to Know about Energy – in School and Society*, London: Falmer Press.

Sousa, D. A. (2001) *How the Brain Learns* (2nd edn), Thousand Oaks, CA: Corwin Press.

Strauss, Anselm and Corbin, Juliet (1998) *Basics of Qualitative Research: techniques and procedures for developing grounded theory*, Thousand Oaks, CA: SAGE Publications.

Sutton, Clive (1974) Language and communication in science lessons. In C. R. Sutton and J. T. Haysom (eds) *The Art of the Science Teacher*, (Science teacher Education Project) Maidenhead: McGraw-Hill, pp. 41–53.

Taber, K. S. (1991) Girl-friendly physics in the national curriculum, *Physics Education*, 26 (4), pp. 221–26.

Taber, K. S. (1994) Misunderstanding the ionic bond, *Education in Chemistry*, 31 (4), pp. 100–03.

Taber, K. S. (1998a) The sharing-out of nuclear attraction: or I can't think about physics in chemistry, *International Journal of Science Education*, 20 (8), pp. 1001–14.

Taber, K. S. (1998b) An alternative conceptual framework from chemistry education, *International Journal of Science Education*, 20 (5), pp. 597–608.

Taber, K. S. (2000a) Finding the optimum level of simplification: the case of teaching about heat and temperature, *Physics Education*, 35 (5), pp. 320–25.

Taber, K. S. (2000b) Case studies and generalisability – grounded theory and research in science education, *International Journal of Science Education*, 22 (5), pp. 469–88.

Taber, K. S. (2000c) Should physics teaching be a research-based activity?, *Physics Education*, 35 (3), pp. 163–68.

Taber, K. S. (2000d) The Chemical Education Research Group Lecture 2000: Molar and molecular conceptions of research into learning chemistry: towards a synthesis, available at www.rsc.org/lap/rsccom/dab/educ002.htm, accessed 1 August 2004.

Taber, K. S. (2001a) The mismatch between assumed prior knowledge and the learner's conceptions: a typology of learning impediments, *Educational Studies*, 27 (2), pp. 159–71.

Taber, K. S. (2001b) Building the structural concepts of chemistry: some considerations from educational research, *Chemistry Education: Research and Practice in Europe*, 2 (2), pp. 123–58, available at www.uoi.gr/cerp/, accessed 1 August 2004.

Taber, K. S. (2002a) The science of physics teaching, *Physics World*, December 2002, 53–54.

Taber, K. S. (2002b) *Chemical Misconceptions – Prevention, Diagnosis and Cure*, London: Royal Society of Chemistry.

Taber, K. S. (2003a) The atom in the chemistry curriculum: fundamental concept, teaching model or epistemological obstacle?, *Foundations of Chemistry*, 5 (1), pp. 43–84.

Taber, K.S. (2003b) Facilitating science learning in the inter-disciplinary matrix – some perspectives on teaching chemistry and physics, *Chemistry Education: Research and Practice*, 4(2), pp.103–14, available at www.uoi.gr/cerp/, accessed 1 August 2004.

Taber, K. S. (2003c) Understanding ionisation energy: physical, chemical and alternative conceptions, *Chemistry Education: Research and Practice*, 4 (2), pp. 149–69, available at www.uoi.gr/cerp/, accessed 1 August 2004.

Taber, K. S. (2003d) Lost without trace or not brought to mind? – a case study of remembering and forgetting of college science, *Chemistry Education: Research and Practice*, 4 (3), pp. 249–77, available at www.uoi.gr/cerp/, accessed 1 August 2004.

Taber, K. S. (2003e) Mediating mental models of metals: acknowledging the priority of the learner's prior learning, *Science Education*, 87, pp. 732–58.

Taber, K. S. (2003f) Examining structure and context – questioning the nature and purpose of summative assessment, *School Science Review*, 85 (311), pp. 35–41.

Taber, K. S. and Watts, Mike (1997) Constructivism and concept learning in chemistry – perspectives from a case study, *Research in Education*, 58, pp. 10–20.

TLRP (2003) Consulting pupils about teaching and learning. Teaching and Learning, Research Briefing No. 5, June 2003, Cambridge: Teaching and Learning Research Programme.

Tobin, Kenneth, Kahle, Jane Butler and Fraser, Barry J. (1990) *Windows into Science Classrooms: problems associated with higher-level cognitive learning*, Basingstoke: Falmer Press.

Topham, T., Holmes, A., Rickwood, B., Unsworth, S. and Chanery, S. (2002) *Applied Science GCSE Double Award Pupils' Book and Teachers' Guide*, London: Hodder Headline.

TTA (2003a) *Qualifying to Teach: professional standards for qualified teacher status and requirements for initial teacher training*, Teacher Training Agency.

TTA (2003b) *Induction Standards for NQTs*, Teacher Training Agency.

UCET (2003) *Higher Education Awards for the Teaching Profession: a discussion paper*, Universities Council for the Education of Teachers Chair's Advisory Group, available at www.ucet.ac.uk/Docs/cagjun03.doc, accessed 1 August 2004.

University of York Science Education Group (1990–2) *Science: The Salters Approach*, Oxford: Heinemann.

University of York Science Education Group (2000a*) Salters Advanced Chemistry Chemical Storylines* (2nd edn), Oxford: Heinemann.

269

University of York Science Education Group (2000b) *Salters Advanced Chemistry Chemical Ideas* (2nd edn) Oxford: Heinemann.

University of York Science Education Group (2000c) *Salters Horners Advanced Physics*, Oxford: Heinemann.

University of York Science Education Group and The Nuffield Foundation (2002) Salters Nuffield Advanced Biology, Trial materials, York: UYSEG and London: Nuffield.

Viennot, L. (1985) Analyzing students' reasoning: tendencies in interpretation, *American Journal of Physics*, 53 (5), May 1985, pp. 432–36.

Vygotsky, Lev (ed. Alex Kozulin) (1986) *Thought and Language*, London: MIT Press (first published in Russian in 1934).

Wall, A. (2002) Using formative assessment to support pupils in setting effective short-term targets, *Education in Science*, 197, pp. 18–19.

Warren, J. W. (1983) Teaching about energy (letter), *Physics Education*, 18, p. 55.

Watson, R. and Fairbrother, R. (1993) Open-ended work in science (OPENS) project: managing investigations in the laboratory, *School Science Review*, 75 (271), pp. 31–38.

Watson, R., Prieto, T. and Dillon, J. (1995) The effect of practical work on students' understanding of combustion, *Journal of Research in Science Teaching*, 32 (5), pp. 487–502.

Watts, M. (1983) A study of schoolchildren's alternative frameworks of the concept of force, *European Journal of Science Education*, 5 (2), pp. 217–30.

Wellington, J. (1989) *Skills and Processes in Science Education: a critical analysis*, London: Routledge.

Wellington, J. and Osborne, J. (2001) *Language and Literacy in Science Education*, Buckingham: Open University Press.

White, J. (2004) *Rethinking the School Curriculum*, London: RoutledgeFalmer.

White, R. and Gunstone, R. (1992) *Probing Understanding*, London: Falmer Press.

Wightman, Thelma, in collaboration with Peter Green and Phil Scott (1986) *The Construction of Meaning and Conceptual Change in Classroom Settings: case studies on the particulate nature of matter*, Leeds: Centre for Studies in Science and Mathematics Education – Children's learning in science project, February 1986.

Wragg, E. C. and Wood, E. K. (1994) Teachers' first encounters with their classes. In B. Moon and A. Shelton Moyes (eds) *Teaching and Learning in the Secondary School*, London: Routledge.

Wray, J. (ed.) (1987) *Science in Process*, London: Heinemann.

Zajano, Nancy C. and Edelsberg, Charles M. (1993) Living and writing the researcher-researched relationship, *International Journal of Qualitative Studies in Education*, 6 (2), pp. 143–57.

Index

action research 226, 239, 242–8
Adey, Philip 232
Advancing Physics 84–6, 156
A level xix, 10, 47, 51, 80–8, 90,
    128, 144, 198, 237
'All our futures' 59
alternative conceptions 35, 102–7,
    131–2, 141–2, 152, 164, 182,
    201, 221
analogies 34, 73, 112, 152, 190,
    191–2, 195
anchoring 89, 139–40
APECS (Able Pupils Experiencing
    Challenging Science) Project 252
Applied Science 74, 79–80, 86–8
appraisal 216
argument 75, 145, 183, 187–8, 205
Armstrong, Henry 42
AS level xix, 80, 82–7
Ausubel, David 124
authentic science education 39, 49,
    65, 144, 147, 155, 204, 212–13,
    242
A2 level xix, 9–10, 82–7, 128
ASE (Association for Science
    Education) xix, 13, 15, 36–7, 62,
    119, 213, 215, 217, 219–20
assessment 57, 78, 88; diagnostic
    201; for learning 201; formative

183, 194, 201–3; informal 204–6;
    peer- 204; planning for 127–8;
    portfolio 79, 87; self- 204;
    summative 198
attainment target xix, 46, 96

BA (British Association for the
    Advancement of Science) xix, 44,
    56, 215
Bachelard, Gaston 30
balanced science 48, 77
Bennett, Judith 221, 239
BERA 249, 251
'Beyond 2000' 52–60, 74–5
big ideas in science 76
biology, not a secondary curriculum
    subject 95–7
Black, Paul 49
Bloom's taxonomy of educational
    objectives 73, 125–6
Bruner, Jerome 242
building blocks as a key idea in
    science 68
'buzz' groups 188

Callaghan, James 48
card games 186–7
CARN 250
career entry profile 216

CASE (Cognitive Acceleration through Science Education) 232–4
case studies 218, 226
Cavendish Laboratory 33
cells as a key idea in science 66
challenging learners 63–4, 72, 125, 131, 175, 183–4, 190, 205, 230, 237, 252
change as a key idea in science 68
Chemical Education Research Group 99
chemical reactions as a key idea in science 68
Chemistry in Context 237
chemistry, not a secondary curriculum subject 95–7
choice 53, 56, 78, 86
citizenship, science for 52
classroom talk 145
Claxton, Guy 167
CliSP (Children's Learning in Science Project) 244–5
cognitive development 231
cognitive dissonance 149, 233
collaborative research 244, 246–7
Collins, Sue 51
compartmentalisation of learning 44, 66
computer-based learning (CBL) 154
computer modelling 156
concentration 146, 169, 188
concept mapping 139–40, 181
conceptual entities of science 142, 189
conceptual fossils 29–31, 143
conceptual readiness 233
concrete operational thinking 231
confidence 174, 176, 223
conservation as a key idea in science 68
constructivism 88–9, 228, 234; research based upon 90, 228–31,

245; science as socially constructed 40–1, 62, 135, 189; social 59, 231, 238; theories of learning 44, 230–1, 240
contemporary issues 53
content analysis 137
content-led science courses 44–7
context 38, 230; based courses 81–4, 87, 237–9; in examination questions 201
continuity 65–8, 122–3
costumes of teachers 160–3, 169–70
coursework 33; *see also* Sc1
CPD (Continuing Professional Development) xix, 63, 68, 107, 207, 213–14
creation science 40
creativity 20, 36, 37, 58–9, 70, 104, 140, 179–80, 189, 199
criterion-referencing 198
critical event 149, 173
critical reflection 212
critical thinking 185–6
CSE (Certificate of Secondary Education) xix, 46, 48, 119
cue cards 164–5
curriculum 5, 115; hidden 115; spiral 139
curriculum analysis taxonomy 232
curriculum model 28–8, 143–4, 153; of scientific enquiry 31–4
curriculum science 21, 35–6, 40, 200, 257

DARTs 140, 146–7, 185
data-logging 155–6
demarcation 95, 98
demonstrations 70, 148–9, 161, 164–6, 172, 182, 184, 230, 234
development plan 216
dialogue 189–94

differentiation 49, 122, 147, 180, 186, 205–6, 255
disciplinary matrix 26–8, 94
disciplinary structure of science 25–7, 98
discovery boards 181
discovery learning 42, 228, 234, 236
discussion 185, 187–8
document review 227
domains of expert teaching knowledge 93, 114
Donaldson, Margaret 232
double award science 48, 51, 58, 79–80
drawing 182

ECLIPSE Project 90
education as a service industry 202
Education in Chemistry 219
Education in Science 15
educational research 5–6, 209–10, 217–21, 243
educational software 153–5
educational technicians, teachers treated as 2, 70, 203
educational technology 151–3
electronic discussion lists 151–2, 217
elicitation of learners' ideas 131, 140–1, 201, 230
elitism 199
endings 194–5
energy as a key idea in science 65
England, national context 8–11
entertaining students 162
enthusiasm 36
EPIC model for CBL 154
episodes, in lessons 33, 108, 120, 183–4
equal opportunities 199
ethnography 227

European Science Education Research Association 222
examinations 30, 32, 48, 51, 73, 78–88, 102, 115, 116, 121, 127, 128, 131, 198–201, 233–4, 240
experiments 28, 148–9, 165, 225, 227, 233
explanations 34, 38, 54, 65–7, 74, 76–7, 94, 102, 135–7, 139, 144, 146, 150, 164, 180, 183–4, 186, 187, 188–95, 234, 236, 255
explanatory stories 54, 59
explanatory vacuum, effect of 144
evidence-based practice 6, 253, 258

Feyerabend, Paul 27
focus groups 226
force and motion, common alternative conception 103–4
forces as a key idea in science 66
formal operational thinking 231–2
framework for teaching science at KS3 63–6
further education 9

Galileo 148
Gatsby Foundation 255
GCE (General Certificate of Education) xix, 86–8
GCSE (General Certificate of Secondary Education) xx, 48, 79–80
gender 45, 47–8, 51, 59, 119
genderalisability of findings 247
generalist versus specialist science provision 41–2, 50–1
general science 45
Gilbert, John 73
GNVQ (General National Vocational Qualification) xx

grade inflation 199
grounded theory 220
GTCE xx, 118, 213
GTTR xx, 97
Gunstone, Richard 180

heuristic approach 42, 234
higher degrees in education 221–3, 253
historical cases 38
historical scientific models 29
Holman, John 74
Hunt, Andrew 74
Huxley, Thomas 42
hypothesis testing 218
hypothetico-deductive method 236

ICT (Information and Communications Technology) 46, 150–6, 165
Ideas-about-Science 56, 74–6
Ideas and Evidence 27, 31, 34, 37, 254–5
imagination 36, 59, 120, 187, 189
individuals, students as 112, 132, 167, 205; teachers as 160–1, 173, 195–6
induction, into the profession 14–15
Inhelder, Bärbel, 231
initiation–response–evaluation 205
Institute of Biology (IoB) 214, 219
Institute of Physics (IoP) 84, 99–100, 156, 214, 219
integrated science 48
intellectual honesty 59, 242
interactive whiteboards 151
interdependence as a key idea in science 66
International Journal of Science Education 219
International Journal of Science and Mathematics Education 219

interviews 226
intuitive physics 104
investigations, 235; genuine 148
investigative skills 27, 31–4
ionic bonding 30–1

Journal of Biological Education 219
Journal of Research in Science Teaching 219
Journal of Science Education and Technology 219
Journal of Science Teacher Education 219

key ideas in school science 64–6
key questions 183
KS3 Strategy see National Strategy
Key Stage 5 xx, 9
knowledge construction xiii, 53, 112, 130, 183, 205, 231
Kuhn, Thomas 26, 35

laboratory coats 170
LAMP project 237
language and learning 145–6
language of science 136–7
league tables 200
learning as knowledge construction 130; key points 131
learning-for-understanding 123–4
learning objectives 183, 197
learning styles 130, 154, 167–9, 184
lecture mode of teaching 129–30, 146, 165, 171, 184
lesson activities 144–5, 177, 183–4
lesson planning 128–8, 134, 206
levels 67, 125–7, 198, 205–6
Lewin, Kurt 243
literacy, scientific 54, 74

maintained schools xviii
mathematics and physics as
  complementary teaching subjects
  95
meaningful learning 46, 56, 122–5,
  131, 140, 147, 222, 251
memory 53, 101, 125, 130, 137–40,
  167; limitations of
  138
mentors 180, 216
meta-analysis 220
metacognition 126, 233
Millar, Robin 43, 73, 74, 231
misconceptions 16, 103–4, 239
models 34, 37–8, 54, 65–6, 71, 73,
  75, 84–6, 104, 112, 135, 142–4,
  150, 152, 153, 155–6, 192, 195,
  210, 255, 258; see also
  curriculum models
moderation of teacher marking 32,
  79
modern science in courses 46, 84
molecular model of matter 65
motivation 83–4, 111, 124, 150,
  154, 187, 200, 202, 220, 222,
  234–9
multiple intelligences 168

narrative 56, 123, 139, 193;
  storylines in Salters Courses 82
National Learning Network 154
National Strategy ('Key Stage 3
  Strategy') 33, 63–73, 215
NC (National Curriculum) xx, 18,
  27, 46, 48, 63, 96, 115–16,
  119–20, 150; constraining
  teaching 73, 120, 148, 205
newspaper articles 185
New Zealand 56
norm-referencing 198
Northern Ireland, national context
  8–11

Nuffield curriculum projects 43, 45,
  57, 74, 81, 83, 234, 237

observation 226
Ofsted xxi
Ogborn, John 84
O level xxi, 46–7
OHP (overhead projector) 153
'Opportunity and Excellence' 79
optimum level of simplification 31
Osborne, Jonathan 51

paradigm 26, 219, 228, 231
paradigm-shift 26
particles as a key idea in science 65
pedagogic knowledge 93, 108, 111,
  114, 141, 158, 179–80
pedagogic learning impediments
  106, 142–4
pedagogy 20, 97
peer-observation 207
peer-tutoring 204
personal philosophy of teaching 4–7
PGCE xxi, 105, 129, 210, 222, 223,
  246, 254–6
philosophy of science 25
physics, not a secondary curriculum
  subject 95–7
Physics Education 219
Piaget, Jean 231–2
plenaries 72
PLON 237, 239
POE (predict, observe, explain) 149,
  182
post-compulsory education 9
practical work 46–7, 147–50, 184,
  198, 234–7
praise 131, 194, 202, 206
preparing to teach 114
prerequisite knowledge 125, 138,
  201
prior knowledge 180, 201

process versus content 41–7
professional, what it means to be
2–4, 70, 91, 116–20
professional body 13
professional identity, of science
teachers xv, 4–5, 13–16, 20, 77,
92, 98, 117, 142, 161, 177, 195,
211, 213, 220, 243
programme of study xxi, 96, 116,
120
progression 65–8, 125–7
prompt questions 182–3
pseudo-questions, teachers questions
as 205

QCA xxi, 10; scheme of work
69–70, 121–2
QTS xxi, 198
qualitative data in education 225–7
quantitative data in education
225–7
questionnaires 226
questions 190, 204–5

reader generalisability 243
reading age 185
reflective practitioner, teacher as
209
reinforcement of learning 139–40,
194, 230
reliability 228
research cycle 244
research in science education 89–90
Research in Science Education 219
research networks 247, 249
research partnerships 246–7
research training for teachers 246,
253–6
robustness of teacher subject
knowledge 100
role(s), teaching as 112, 142, 177
rote learning 46, 53, 123, 140

Royal Society of Chemistry (RSC)
13–14, 99, 214, 219, 230
Rutherford 33

safety 170
Salters Institute projects 74, 81–4,
90, 123, 237–9
SAT xxi, 63
SATIS (Science and Technology in
Society) 237
schemes of work 69–70, 121–2
Schools–University Partnership for
Educational Research (SUPER)
247–8
SCIENCE xviii, 1, 20; disciplinary
structure 25–7; nature of 135–6,
175; unity of 21–4
Science Education 219
science education, as a contested
field 212; role of 17
Science Enhancement Programme
255
science for all 41, 48–51
Science in Process 43, 235
Science Learning Centres 221
Science Reasoning Tasks 232–3
scientific attitude 36, 62, 236
scientific enquiry 27–8, 31–4, 235
scientific method 27–8, 236
SCISP 48
school science 16, 20, 143; as
politically determined 95–9, 200,
212; as something other than
science xv, 1, 62, 211–12
School Science Review 15, 219
Scotland, national context 8–11
Sc1 27, 31–4, 96, 148, 210, 235,
254
seating plans 166
Shayer, Michael 232
simulations 153
single science 119

sixth form xxi, 10
social science, education as 4, 209–10, 219; research methodologies 220, 225–7, 246
specialism, teaching outside 16, 77, 94–5, 99
specifications, examination 10
specific learning difficulties 201
stage, teaching room/laboratory as 172–3; theories of development 231–2
stances of students 167–8
starters 72
story telling 192–3
structure in material to be taught 135
student-led activities 187
Studies in Science Education 219
subject knowledge 21–5, 92–4, 100–8, 162; breadth needed for teaching 94, 142–3; QTS requirements 105
Supporting Physics Teaching 11–14 100
surveys 218, 227

teacher assessment 32
teacher-researchers 243
teaching 111, 130, 197; as acting 156; as a craft 163; and lecturing 129–30
technology 53, 56–7, 84
texts in science 147
Thinking Science 233–4

three-part lesson 72
TTA xxii
trainee xviii
transferable skills 43–4, 46, 236
transforming the curriculum 68, 108, 111, 115, 138, 200, 212
translation in planning teaching 137
'Twenty First Century Science' 57, 74–8

University of York Science Education Group 57, 81

validity 228
values 36
variety 71, 130, 184, 237
VCE xxii, 79
VGCSE 58
Viennot, Laurence 104
vocabulary, scientific 185
vocation, teaching as 176
vocational courses 57–8, 79, 86–8
voice 171, student 203

Wales, national context 8–11
Warwick Process Science 43, 235
Wellcome Trust 74, 221
Wellington, Jerry 46
Western science 59–60
White, Richard 180
work experience 80
writing frames 146, 186
writing in lessons 145–6